Customer Choice:

Purchasing Energy in a Deregulated Market

CUSTOMER CHOICE:

PURCHASING ENERGY IN A DEREGULATED MARKET

COMPLIED AND EDITED BY
ALBERT THUMANN, P.E., C.E.M

1999

Published by
THE FAIRMONT PRESS, INC.
700 Indian Trail
Lilburn, GA 30047

Library of Congress Cataloging-in-Publication Data

Customer choice : purchasing energy in a deregulated market / compiled and edited by Albert Thumann.
    p. cm.
Many of the chapters were originally presented at the World Energy Engineering Congress and conferences sponsored by the Association of Energy Engineers.
    Includes bibliographical references and index.
    ISBN 0-88173-307-5
    1. Energy industries--Congresses. 2. Power resources--Purchasing--
        Congresses. 3. Electric utilities--Deregulation--Congresses. 4. Gas
        industry--Deregulation--Congresses. I. Thumann, Albert.
    HD9502.A2C87   1998   333.79--dc21   98-41200
                                                                        CIP

Customer choice : purchasing energy in a deregulated market / compiled and edited by Albert Thumann.

Published by The Fairmont Press, Inc.
700 Indian Trail
Lilburn, GA 30047

Printed in the United States of America

10 9 8 7 6 5 4 3 2 1

ISBN 0-88173-307-5 FP

ISBN 0-13-083820-9 PH

While every effort is made to provide dependable information, the publisher, authors, and editors cannot be held responsible for any errors or omissions.

Distributed by Prentice Hall PTR
Prentice-Hall, Inc.
A Simon & Schuster Company
Upper Saddle River, NJ 07458

Prentice-Hall International (UK) Limited, London
Prentice-Hall of Australia Pty. Limited, Sydney
Prentice-Hall Canada Inc., Toronto
Prentice-Hall Hispanoamericana, S.A., Mexico
Prentice-Hall of India Private Limited, New Delhi
Prentice-Hall of Japan, Inc., Tokyo
Simon & Schuster Asia Pte. Ltd., Singapore
Editora Prentice-Hall do Brasil, Ltda., Rio de Janeiro

JK

TABLE OF CONTENTS

ACKNOWLEDGMENTS

The author wishes to express appreciation for the numerous contributors to this work. Many of the chapters were originally presented at the World Energy Engineering Congress and conferences sponsored by the Association of Energy Engineers. In addition, this reference includes papers contributed to *Strategic Planning for Energy and the Environment* and the *Cogeneration and Competitive Power Journal*, and *Energy Engineering Journal*.

Hopefully, by including these contributions in this source book, these works will gain wider exposure.

INTRODUCTION

The energy purchasing industry is rapidly changing as new state regulations go into effect. While developing this book it became apparent that there was no reference source addressing the myriad of questions energy users were facing in the energy procurement arena. Even though the material contained was current at the time of writing, the author advises the reader that regulation information is bound to change rapidly.

There is no question that there is great opportunity for individuals to learn how to buy and sell power in today's marketplace. Individuals who develop this expertise are sure to find career opportunities as deregulation takes hold.

Hopefully, this book will provide a source of information to get started.

CHAPTER 1

THE ROLE OF THE ENERGY PROCUREMENT PROFESSIONAL

E nergy delivered to industrial, commercial, institutional and residential customers costs $270 billion annually. In a deregulated marketplace new opportunities to purchase electricity and gas have emerged.

As we enter the next millennium the energy procurement professional will play a vital role as either a buyer or seller of energy.

For companies to stay competitive in a deregulated world, an understanding is needed on how to buy electricity and gas under the new rules of the game. New businesses are emerging which provide new procurement services to companies. The energy procurement professional must analyze these options and determine the associated risks and savings.

The seller of energy needs to understand the needs of the customer and what options to offer. The stakes are high as this new industry evolves.

As with any new industry new companies are emerging. The energy procurement professional must understand the new companies and entities which are part of this industry. The following is a sampling of the terms associated with the energy procurement industry:

Aggregator: A company that combines end user loads into a group to achieve the best electric price from a power supplier.

Marketer: A company that purchases electricity and or gas from traditional utilities or other suppliers and then resells these services to end users. A marketer takes title to the power.

Power Pool: An independent organization which serves as a short-term spot market where electricity buyers and sellers conduct transactions. This independent organization integrates, coordinates and balances power and consumption by competitive bid.

Power Exchange: The power exchange is the name of a new entity in California that establishes competitive spot market prices for electricity through day and hour ahead auction of demand bids and generation.

Independent Power Producer: A company that generates power but does not have distribution or transmission facilities.

Independent System Operator (ISO): The ISO is a neutral operator who maintains the workings of the electric grid. The ISO controls the dispatch of flexible power plants to ensure that loads match resources available to the system.

Broker: A company or individual who matches the electricity sellers with buyers. A broker does not take title to the power.

Exempt Wholesale Generator (EWG): EWG is an entity which sells power exclusively to other power producers in the wholesale market. EWGs were created under the Energy Policy Act of 1992.

In addition there are other players who have emerged in this new marketplace including unregulated utility subsidiaries, mega-wholesalers and energy service companies (ESCO's).

To gain a perspective on the dynamic forces shaping this industry and profession the Association of Energy Engineers has implemented a comprehensive 1998 survey of its members. Based on 911 responses the following results were found:

1. If you are an enduser, what is the estimated savings that you (and/or your company) are projecting which will be attributable to purchasing electricity and gas in a deregulated environment?

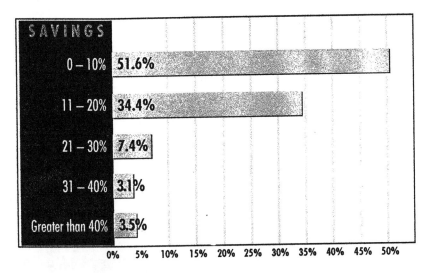

2. Customer choice of power suppliers will lead to:
   Lower electric rates in a marked way: 61.5%
   Higher peak rates: 41.1%
   Customer doing business with new power suppliers: 74.9%

3. Would you use a power marketer or power broker to get the best energy price?
   Yes: 71.5%
   No: 28.5%

4. As a result of customer choice of power supplier, my company is delaying purchasing energy efficient equipment.
    Yes: 11.3%
    No: 88.7%

5. Real time pricing will encourage installing the following *(respondents were told to mark a 4 for the most important and on down to a 1 for the least important factor):*

	Least Important 1	2	3	Most Important 4
Gas cooling equipment	38.8%	26.9%	16.2%	18.1%
Thermal energy storage	32.2%	32.7%	20.3%	14.7%
Energy management systems	14.7%	15.3%	33.8%	36.2%
Metering	17.3%	15.1%	28.2%	39 4%

6. How do you rate your existing utility?

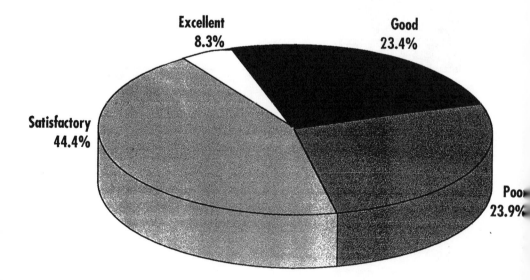

7. How do you rate your utility to customer service?

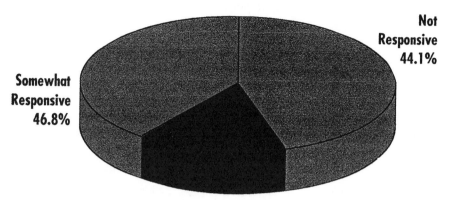

Somewhat Responsive 46.8%

Not Responsive 44.1%

Responsive 9.1%

8. In selecting an energy service provider, indicate the most important programs to be provided (*respondents were told to mark a 9 for the most important and on down to a 1 for the least important factor*):

	Least Important						Most Important		
	1	2	3	4	5	6	7	8	9
Purchasing electricity	7.8%	2.2%	2.4%	2.5%	2.9%	6.3%	8.3%	12.2%	55.4%
Purchasing gas	8.4%	7.7%	4.6%	3.6%	9.2%	9.1%	11.8%	28.4%	17.3%
Project financing	11.4%	10.3%	11.1%	9.7%	14.7%	13.5%	14.0%	8.0%	7.4%
Performance contracting	8.2%	9.1%	9.8%	9.9%	15.6%	13.7%	11.5%	10.4%	11.8%
Facility outsourcing	16.3%	14.5%	15.7%	12.1%	14.2%	9.3%	9.2%	5.3%	3.4%
Improving power quality	5.2%	7.4%	8.9%	12.6%	14.6%	12.0%	18.2%	12.8%	8.4%
Upgrading HVAC & lighting systems	6.2%	5.2%	9.4%	9.8%	14.1%	11.5%	19.6%	12.8%	11.5%
Providing cogeneration alternatives	17.7%	15.7%	11.0%	11.8%	14.1%	9.1%	9.0%	5.2%	6.4%
Integrating billing for all for energy, telecommunications and security	23.7%	10.4%	9.3%	8.1%	11.0%	8.6%	10.4%	8.4%	10.0%

9. How do you select an energy provider (*respondents were told to mark a 6 for the most important and on down to a 1 for the least important factor*)?

	Least Important				Most Important	
	1	2	3	4	5	6
Name recognition	28.4%	21.5%	14.3%	15.1%	12.1%	8.6%
Price of energy	5.6%	2.9%	6.5%	12.8%	23.2%	49.0%
Energy services offered	4.5%	8.1%	14.7%	20.7%	27.2%	24.9%
Location of provider	24.6%	20.6%	17.8%	15.3%	12.4%	9.3%
Perception of "value"	8.7%	12.6%	15.7%	19.5%	22.5%	21.0%
Customer service	2.7%	6.2%	15.1%	22.1%	27.8%	26.2%

10. Will utility mergers improve or hurt the energy industry?
    Improve: 64%
    Hurt: 36%

11. Are you presently involved in energy buying decisions?
    Yes: 57.5%
    No: 42.5%

12. Do you see your role expanding to include energy buying decisions?
    Yes: 64.9%
    No: 35.1%

13. Will lack of new generating facilities and decommission of nuclear power plants lead to power shortages in your area?
    Yes: 31.4%
    No: 68.6%

14. Which services do you believe utilities or their affiliates should provide?

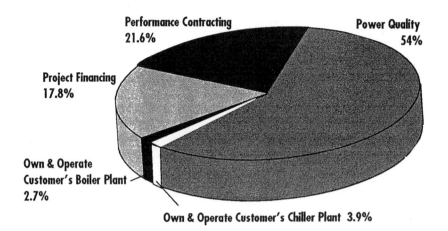

Performance Contracting 21.6%

Power Quality 54%

Project Financing 17.8%

Own & Operate Customer's Boiler Plant 2.7%

Own & Operate Customer's Chiller Plant 3.9%

15. Utilities have made their customers aware of how restructuring will impact their customers.
    Agree: 32.2%
    Disagree: 67.8%

CERTIFIED ENERGY PROCUREMENT (CEP) PROGRAM

With the goal of raising the professional standards in the fields involved with the purchasing, selling, and marketing of electricity and natural gas, the Association of Energy Engineers (AEE) has established the Certified Energy Procurement Program for professional certification. Since 1981, AEE has certified more than 4,000 professionals within the energy industry. AEE's certification programs are recognized by governmental agencies, including the U.S. Department of Energy and the U.S. Agency for International Development, as well as by utilities, end users, and energy service companies.

When professionals earn the right to put the initials "CEP" behind their name, they are distinguishing themselves among those involved professionally in today's restructured energy marketplace. They have demonstrated high levels of experience, competence, and specialized knowledge within their field.

Requirements to Sit for the CEP Exam

Each applicant for CEP professional certification must attend AEE's three-day "Fundamentals of Buying & Selling Energy" training program and complete and pass a 4-hour written exam, as well as meet the following criteria.

The candidate must have:

1. A 4-year degree from an accredited university or college, in science, engineering, architecture, business, law, finance, or related field, or be a registered Professional Engineer (P.E.) or Registered Architect (R.A.). In addition, the applicant must have at least three years of experience in energy or building facility management, or real estate, or procurement, or brokering;

or

2. A 2-year technical degree or a 4-year non-technical degree, with five years experience in energy or building or facility management, or real estate, or procurement, or brokering;

or

3. Ten years of experience in energy or building or facility management, or real estate, or procurement, or brokering;

or

4. The current status of Certified Energy Manager (CEM).

Format of the CEP Exam

The four-hour CEP exam is given in conjunction with the Fundamentals of Buying & Selling Energy 3-day training program. The examination questions are based on concepts and experiences basic to purchasing, selling, and marketing electricity and natural gas. The exam is open book, and the questions are a mixture of multiple choice and true or false.

THE FUNDAMENTS OF BUYING & SELLING
ENERGY PREPARATORY COURSE

In order to help professionals prepare for the CEP examination the Association of Energy Engineers has developed the following course. For course dates and details on the CEP program visit AEE's web site at www.**acecenter.org** or FAX at (770)-446-3969 or call (770)-447-5083.

This comprehensive 3-day instructional program has been designed to provide the specific training and background information needed by professionals preparing to sit for the Certified Energy Procurement (CEP) Professional examination. As energy restructuring continues to unfold nationwide, there are an increasing number of individuals who have developed specialized expertise, both on the purchasing side and on the selling side of the marketplace, as well as a growing need for this kind of special knowledge. AEE's new CEP professional certification provides a defined standard against which this expertise may be established and demonstrated.

The program covers the full spectrum of topics essential to the energy procurement process, covering both electricity and natural gas, from both a purchasing/procurement and a selling/marketing perspective. You'll learn the best approaches for defining corporate energy goals, assessing energy use parameters, evaluating energy provider options, comparing competitive proposals, and integrating the benefits of competitive energy procurement into an overall energy management program. On the selling side, the program will examine proven strategies for building a customer base, answering RFPs, and establishing overall sound business practices that will build your market share and status as a player in the marketplace. The growing field of energy trading will also be explored, covering futures, options and derivatives, hedging, and effective risk management.

COURSE OUTLINE

**Legislation, Regulation, and
Energy Outlook for Energy Procurement**
- Energy Usage Breakdown
- History of Energy Management
- Importance of Energy Procurement

**The Structure of the
Electric Utility Industry**
- Generation, Transmission, and Distribution
- Review of the Physical System
- Billing Information
- Traditional Ways of Seeking Lower Rates

Purchasing Electricity under the New Rules
- Procurement Process
- Information Gathering
- Aggregation
- Creating the RFP
- Selecting a Power Marketer, Broker, or Aggregator
- Evaluating Bids
- Negotiating Contracts
- Contract Adherence

The Structure of the Natural Gas Industry
- Producer, Transporter, and Distributor
- Review of the Physical System
- Current Rate System
- Billing Information
- Traditional Ways of Seeking Lower Rates

Purchasing Natural Gas under the New Rules
- Procurement Process
- Information Gathering

- Aggregation
- Creating the RFP
- Selecting a Gas Marketer, Broker, or Aggregator
- Evaluating Bids
- Negotiating Contracts
- Contract Adherence

Metering, Load Profiling, and Real Time Pricing
- Metering
- Load Management
- Real Time Pricing
- Fuel Switching

Energy Trading and Risk Management
- NYMEX, Definitions
- Futures, Options, and Derivatives
- Hedging
- Risk Management
- Portfolio Management

Fundamentals of Gas and Electric Marketing
- Gas Producer, Transporter, and Distributor
- Power Generator, Transporter, and Distributor
- LDC Tariffs
- Answering RFPs
- Customer Relations
- Account Management

Energy Cost Avoidance Strategies
- Cogeneration
- On-Site Power Generation
- Gas Cooling
- Energy Management

The 4-hour CEP exam is administered at the close of instruction on day three.

CHAPTER 2

RETAIL COMPETITION—
STATE-BY-STATE UPDATE

Editor's Note: This update was compiled in February 1998. Due to fluidity of the marketplace, this update will change rapidly as deregulation takes hold across the nation. Hopefully, this chapter will give a snapshot of the deregulation process as it is occurring. The chapter is based on articles published in the February and March 1998 issues of *Energy User News*. Reprinted with permission *Energy User News*.

Attempting to publish a state-by-state analysis of electricity restructuring activities is a little like shoveling a sidewalk while it's still snowing—you know more is coming, but you have to do what you can before you're overwhelmed and buried.

In the electricity industry, it's a blizzard. Each day brings another legislative decision, another report, another end user eager to receive power at a lower cost from an **alternative** power supplier. It's time consuming just trying to keep up with what's happening in one state, let alone the entire country. For these reasons, EUN is publishing this two-part report on the most current deregulation news in all 50 states.

As this report shows, restructuring is happening at vastly different speeds across the nation. Some states are on the brink of

---

By Richard Zomnir, president, Strategic Energy Ltd., Pittsburgh, an energy consulting and management firm to end users of electricity and natural gas.

full completion, while others are just beginning to study the possibilities of introducing choice.

While the rate of adoption differs, the issues are often the same; terms such as stranded costs, pilot programs, and aggregations surface in nearly every listing.

Above all, this report shows one thing: electricity restructuring is gathering momentum. Hopefully, this special report will help end users prepare for open access by providing a firm overview of issues and high-lighting what to look forward to in the future.

The round-up was compiled and written by Richard M. Zomnir, president of the energy consulting and management firm, Strategic Energy Ltd., Pittsburgh, with revisions by EUN as needed to update specific state programs and initiatives.

THE ABCs OF DEREGULATION (by Grant Gegwich)*

These common terms and acronyms often appear in discussions of electricity deregulation across the nation. This guide will help you decipher the industry jargon.

Access charges
Fees charged independent producers by the owner of a transmission or distribution network.

Aggregator
Any entity that combines the loads of end users into a group to achieve the best rates from an electric supplier.

Ancillary services
Any extra services provided by a power provider.

*Material for this glossary was obtained from the Public Forum Institute, Building Owners and Managers Association International, and the California Public Utilities Commission.

APPA

American Public Power Association. A national association representing municipally owned and other publicly owned electric utilities.

Broker

The company or person that matches electricity buyers and sellers. Unlike a marketer, a broker does not take title to the power.

Direct access

The ability to purchase electricity directly from the wholesale market rather than through a supplier.

Distribution

The delivery of electricity to an end user through low voltage distribution lines.

EPACT

The Energy Policy Act of 1992. Legislation that created exempt wholesale generators and gave FERC the authority to order and condition access by eligible parties to the interconnected transmission grid; also set energy efficiency standards for certain products.

EWG

Exempt wholesale generator. An entity that sells power exclusively to other power producers in the wholesale market, but not to end users directly. EWGs were created under the Energy Policy Act of 1992 and are exempt from the Public Utility Holding Company Act.

FERC

Federal Energy Regulatory Commission. Is responsible for regulating the price, terms, and conditions of transactions in the U.S. wholesale electricity market. Also handles intrastate electricity issues.

Generation
The act of turning various forms of energy input into electrical power. The portion of electrical service that is open to competition.

Grid
The transmission and distribution networks operated by electrical utilities.

IOU
Investor-owned utility. These entities are taxed like other businesses but are regulated by the state and federal governments.

IPP
Independent Power Producer. An entity that produces power but does not have transmission or distribution facilities.

ISO
Independent system operator. A neutral operator responsible for maintaining the workings of the electric grid. The ISO performs its function by controlling the dispatch of flexible plants to ensure that loads match resources available to the system.

Marginal cost
The cost to a utility to provide a kilowatt hour of electricity.

Marketer
Any entity that buys electric energy, transmission, and other services from traditional utilities or other suppliers, and then resells those services to end users.

NARUC
National Association of Regulatory Utility Commissioners. An advisory council of governmental agencies that regulates utilities and carriers.

Order 888

FERC rule that promotes wholesale competition through open access, non-discriminatory transmission services by public utilities.

POU

Publicly owned utility.

Power pool

Independent organization that coordinates, integrates, and balances power production and consumption by competitive bid. Serves as a short-term spot market where electricity buyers and sellers can conduct transactions.

PUC

Public Utilities Commission. Name of a state agency that regulates intrastate electricity transactions and retail electric service. It must abide by FERC guidelines. Also commonly known as Public Service Commissions.

PUHCA

The Public Utility Holding Company Act of 1935. Prohibits acquisition of any wholesale or retail electric business through a holding company unless that business forms part of an integrated public utility system when combined with the utility's other electric business. Also restricts ownership of an electric business by non-utility corporations.

PURPA

Federal legislation enacted in 1973 that requires utilities to buy electric power from "qualifying facilities" at an avoided cost rate, which is equivalent to what it would have cost the utility to generate or purchase that power itself.

PX

Power Exchange. The name of the a new entity in California that will establish a competitive spot market for electric power through day and hour ahead auction of generation and demand bids.

Reliability

Refers to the capability of an energy supplier to provide a necessary load at all times, as well as the ability of the electric system to withstand sudden disturbances, such as electric short circuits or loss of system facilities.

Retail wheeling

The ability of an end user to purchase electricity from a supplier of choice and transmit it over the transmission grid.

Standard offer

The rates a utility intends to charge end users under competition.

Stranded costs

Revenues and assets utilities expect to lose from a transition to a deregulated marketplace.

Tariff

A schedule of rates, terms, and conditions for service that utilities are required to file with the state public utility commission.

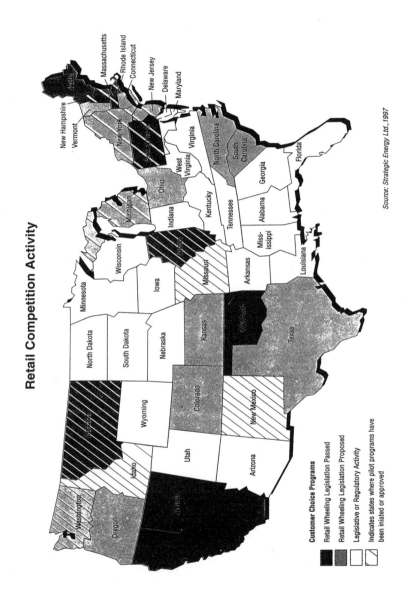

Retail Competition Activity

Customer Choice Programs

- Retail Wheeling Legislation Passed
- Retail Wheeling Legislation Proposed
- Legislative or Regulatory Activity
- Indicates states where pilot programs have been inlated or approved

Source: Strategic Energy Ltd., 1997

Alabama: The Alabama Electricity Consumers Coalition and American Energy Solutions filed suit in U.S. District Court Jan. 27,1997, challenging the constitutionality of a law passed in 1996 that guaranteed stranded cost recovery to utilities. The groups claim that the law violates the Commerce Clause of the Constitution because it prohibits the development of competitive interstate trade. The lawsuit also states that applying the law blocks the sale of energy from independent power producers to new customers, contradicting statutes created in the 1978 Public Utility Regulatory Policies Act. If the lawsuit is successful it could be the first time in the nation that stranded cost legislation is overturned by the courts using the Commerce Clause. The 1996 bill that gave the state's utilities full stranded cost recovery was signed May 6, 1996, by Governor Fob James and was supported by Alabama Power, Birmingham, and other utilities.

Alaska: Two electricity competition bills were proposed in the state's 1997 legislative session, but were rolled over to the 1998 session due to inactivity. One bill proposed allowing retail competition in the Anchorage area; the other wanted to slow down competition by requiring the Public Utilities Commission to find "clear and convincing" evidence of the need for competition. While the second bill may seem contrary to developments in other states, most of Alaska is served by isolated municipal electric companies and cooperatives for which retail wheeling could not apply.

Arizona: The Arizona Corporation Commission announced in January 1997 that it would not grant a rehearing to utilities challenging the restructuring plan it developed in December 1996. The decision by the commission caused two of the state's largest investor-owned utilities, Arizona Public Service, Phoenix, and Tucson Electric Power, Tucson, to file an appeal March 2, 1997. The utilities' concerns revolve around guaranteed stranded cost recovery and system reliability, as well as challenging the authority of the commission to enforce the deregulation plan. This case has not been resolved. Meanwhile, Arizona's state legislature has held hearings in preparation for proposed legislation that was scheduled for January 1998. There is some indication that the legislated plan will be less complicated than the commission's, such as allowing all customers access on a single date rather than through a phase-in process.

The commission's December 1996 plan approved an electricity industry restructuring plan that will phase in retail access over four years, with 20 percent of the load open to competition by 1999, 50 percent by 2001, and 100 percent by 2003. The plan required the state's utilities to submit restructuring plans by the end of 1997. Stranded cost recovery will be allowed under an exit fee arrangement, but mitigation measures will be required.

Arkansas: In December 1997 the Arkansas Public Service Commission approved an electric restructuring plan by Entergy, New Orleans. Sources say end user groups are pleased with the plan because it offers double-digit rate reduction for small commercial users and a transition cost account so stranded costs are not handled only by smaller customers. The proposed $160 million in rate reductions would begin in 1998 and continue into 1999.

The Arkansas Electric Energy Consumers, a group of large industrial users, had filed a docket with the commission July 31, 1997, asking state regulators to reject Entergy's proposal. According to the group, Entergy's plan promotes monopoly domination by allowing the company to recover $784 million in stranded costs that it never proved would exist during the seven-year transition period. The group claims these costs ensure that Entergy will emerge after the seven years in a dominant position over its competitors. In response to the filing, Entergy revised its original proposal by suggesting a review of its stranded cost position three years after its plan is initiated.

The Arkansas legislature also passed a law April 1, 1997, that created a joint committee to study retail competition. The committee is required to complete its report by January 1999.

California: The opening of a competitive electricity market, which was scheduled to take place Jan. 1, 1998, was delayed Dec. 22, 1997, because the computer systems at the Independent System Operator (ISO) and Power Exchange (PX), the two entities that will coordinate buying and selling of power in the state, were not ready. The new target date was set for March 31, 1998.

The California Public Utilities Commission voted Dec. 16, 1997, to allow utility affiliates to compete within their parent companies' service territory and use their parent companies' logos and names in their marketing efforts, with some conditions. All advertisements of affiliates must include a disclaimer that indicates the affiliate marketer is not regulated by the commission.

On Oct. 8, 1997, Governor Pete Wilson signed a bill that requires electricity retailers to inform end users about the source of the electricity they are selling. At a minimum, power providers must describe to end users what sources comprise the statewide power mix. Providers claiming to sell power from certain sources must be able to verify that information.

In September 1997, power marketers won a small victory over investor-owned utilities when regulators ruled in their favor on the calculation of competitive transition charges. The Public Utilities Commission adopted a charge based on an average monthly energy cost, as opposed to one calculated on an hourly basis. This order is a modification of the commission's recent approval of the top-down approach to stranded cost recovery. According to this new plan, the power exchange price will be subtracted from the existing frozen rates—calculated on a monthly basis and based on customer class averages—with the remainder paid to the existing utility. The rate paid to the utility, which includes stranded cost recovery, will fluctuate hourly, inverse to the fluctuations of the power exchange.

On May 6, 1997, the commission unanimously voted to grant immediate access to end users wishing to select an alternative supplier beginning Jan. 1, 1998 (now moved to March 31). Customers began selecting alternative suppliers in November 1997. The ruling eliminated a four year phase-in period that was outlined in the original plan. Also, part of the vote was a stipulation that customers with loads over 20 kilowatts (kW) must install hourly metering equipment at their own expense if they wish to participate in direct access. End users with loads under 20 kW will be able to use load profiling or advanced metering.

The Federal Energy Regulatory Commission (FERC) declared Dec. 18, 1996, that California's investor-owned utilities have more market power than they admitted, thus denying them the right to sell power into the Western Power Exchange at market rates. FERC studies show that even though Pacific Gas & Electric (PG&E), San Francisco, and Southern California Edison (SCE), San Dimas, Calif., offered to divest portions of their capacity through the sale of fossil-fuel plants, the utilities would still have enough power to manipulate the state's electric market. On Nov. 26,1996, FERC gave conditional approval to a plan filed by the state's three major utilities to form the ISO and a separate statewide PX.

Colorado: Following the rejection of five previous restructuring bills, the state legislature abandoned a sixth initiative in May 1997. The bill, introduced by Senator David Wattenburg, would have established an advisory panel consisting of 26 stockholders to evaluate the most beneficial process to bring retail choice to end users through public hearings and the evaluation of other states' deregulation programs. The bill was rejected because of the projected cost of implementation.

Connecticut: Despite the initial optimism surrounding a proposed bill to bring retail choice to Connecticut rate payers, the legislature dropped its attempt to pass a measure in 1997. The House had been considering a bill that came out of committee in March, but attempts at passage in the state's Senate fell apart at the last minute. The failure was due to a lack of consensus on issues surrounding stranded costs for nuclear facilities and the removal of a guaranteed 10 percent reduction in rates for consumers. Not wanting to fall too far behind the other New England states, legislators are expected to revive the bill in early 1998.

Delaware: A report issued by the Delaware Public Service Commission suggested that the state phase in retail access between April 1999 and April 2003, taking place in equal increments of 25 percent of the utilities' generation load each year during the four-year time frame. The commission, however, is now recommending that competition be introduced over three years.

The commission suggests that utilities be required to submit a plan for unbundling to the Public Service Commission within 60 days of a deregulation law being enacted. Six months later, unbundled rates would go into effect. During the three-year transition period, stranded costs would be covered by a market transition charge developed by the commission. If actual transmission costs exceed revenues generated by the charge by 25 percent or more, the charge could be extended. Under the plan, utilities would be required to file a report estimating their stranded costs to the commission three months after a law is enacted. The commission was scheduled to make recommendations to Delaware's General Assembly and the governor in January 1998.

District of Columbia: Continuing a restructuring investigation started last year, the District of Columbia Public Service Commission issued a notice of inquiry that listed specific questions it would investigate concerning retail competition. Answers to the questions were received from eight stakeholders, and responses will be developed by the commission over the next few months. No schedule for a draft or final ruling has been given.

Florida: A legislative committee recently rejected an effort to get the Florida Public Service Commission to study retail wheeling. This rejection follows defeats in 1996 of measures to introduce direct access in a limited form. Responding to intense lobbying by Florida utilities, the committee said studying retail wheeling is tantamount to implementing it, leaving the Florida commission one of the few in the nation that has not investigated retail competition. The commission, however, is informally tracking developments in other states.

Georgia: Despite a lack of legislative action on retail electricity competition, the Georgia Public Service Commission has been holding a series of informal workshops on retail restructuring issues. Some issues discussed include the status of developments in other states, economic benefits to Georgia consumers, and stranded costs. The workshops also have been considering whether existing competitive programs in the state are adequate. Existing state laws give new industrial customers with over 900 kilowatts of load a one-time choice among state suppliers of electricity. There are, however, relatively few suppliers to choose from in Georgia.

Hawaii: Although retail competition can have only a limited meaning in Hawaii, the state's Public Utilities Commission has begun to develop a draft restructuring plan and a formal investigation into the issues.

Idaho: On Feb. 28, 1997, Idaho passed a state law that requires a legislative committee to develop recommendations and legislation for retail competition by 1998 so a bill could be considered in the 1999 session. This legislation was passed in response to recent state and federal activity that may impact the state's already low rates. A more aggressive bill that could have implemented competition by Jan. 1, 1998, was shelved.

In early September 1996, the Public Utility Commission approved a retail wheeling pilot program volunteered by Washington Water Power, Spokane, Wash., which serves some load in the state. As with the plan filed in Washington, the utility's largest end users are able to choose competing suppliers for up to one-third of the total load.

On Dec. 5, 1996, a steering committee formed by the governors of Idaho, Montana, Oregon, and Washington finalized an electricity restructuring plan that aimed to keep low-cost power in the region. The Northwest Energy System Comprehensive Review, formed in January 1996, focused on the role of the Bonneville Power Administration in a competitive market and repayment of that agency's large debt. While the group agreed on a "subscription" system that will eliminate this debt by 2015, the plan also called for some degree of customer choice by July 1, 1999.

Illinois: After passing by an overwhelming margin in the Illinois House of Representatives, an electricity restructuring bill, complete with a 20 percent rate cut, was signed into law by Governor Jim Edgar in December 1997.

In an attempt to avoid the Governor's veto power, the House added a trailer to make the bill more palatable to low-cost utility Central Illinois Light Co. (Cilco), Peoria, Ill. Cilco had long argued that the bill unfairly penalizes their company, which has rates far below other utilities in the state.

Cilco's rates, for example, are about 40 percent lower than Commonwealth Edison in Chicago. To appease the utility, the bill's trailer allowed Cilco to reduce residential rates by 2 percent in 1998, another 2 percent in 2000, and 1 percent two years later. Under the original plan, the utility would have been forced to implement an immediate 5 percent rate cut, with more to come later.

With the signing of this bill, commercial and large industrial end users in Illinois will be able to choose their electricity supplier in 1999. Stranded costs will be recovered through an additional charge on rate payers' bills through 2006.

With its pilot programs, Illinois was the first state in which end users could openly shop the competitive electricity marketplace, albeit on a limited basis (see January 1998 EUN, page 1). The retail wheeling experimental program proposed by Illinois Power, Decatur, Ill., began April 25, 1996, and one by Cilco began April 30, 1996.

Illinois Power allows 21 of its largest industrial end users to purchase 100 percent load factor blocks of power from two megawatts (MW) to 10 MW from off-system suppliers. The program uses Illinois Power's open access transmission rate filed with FERC and includes no stranded cost recovery, although Illinois Power reserves the right to collect stranded costs resulting from the experiment at a later date. Aggressive response to the program has resulted in delivered, bundled rates reportedly as low as two cents per kilowatt hour (kWh) for some end users.

Cilco developed two retail wheeling pilot program tariffs: one for industrial customers, and one for commercial and residential customers. Under Rate 33, industrial loads over 10 MW, which includes Cilco's eight largest customers, have access to off-system suppliers for up to a total of 50 MW of load. Under Rate 34, commercial and residential customers in three designated "open access sites" are able to purchase power from alternative suppliers. Like the Illinois Power plan, no stranded costs are recovered. Results show prices delivered to the Cilco control areas as low as 1.6 cents per kWh; distribution charges would then be added.

Indiana: The Indiana General Assembly's Regulatory Flexibility Study Committee, which is charged with studying whether the state should pursue electricity competition, concluded that it may be best to wait at least another year. The committee took no position for or against deregulation, citing an abundance of "unanswered questions" regarding the effects a competitive market would have on the state.

On May 6, 1997, Indiana's governor signed a bill that required the existing legislative committee to study deregulation and, if appropriate, recommend legislation. This law resulted from investor-owned utilities and consumer advocacy groups who opposed a bill submitted to the Indiana legislature Jan. 16, 1997, sponsored by a group of investor-owned utilities and manufacturing groups. This relatively aggressive plan outlined a structure that would initiate choice for all consumers beginning Oct. 1, 1999. All end users selecting to choose an electric supplier that is not their native utility would be subject to a market access charge. This charge is the difference between the state average electricity rate, 3.7 cents per kWh, and the current rate the native utility charges.

Iowa: On Feb. 10, 1997, the Iowa Utilities Board issued a report on electricity industry restructuring after two years of study.

The report proposed a "wait-and-see" approach and recommended that the board continue to monitor developments in other states. The next step, according to the report, would be to develop an implementation plan, but no timetable was suggested. Legislative hearings have been held to review restructuring bills in parallel with the board's work. No legislative proposals are expected until the 1998 session.

Kansas: The word from the governor's office in Kansas is that deregulation "absolutely will happen," but at a slow pace. According to Mike Matson, press secretary for Governor Bill Graves, "It's nice to be first, but it's better to be best."

A draft bill that would bring retail choice to consumers by July 1, 2001, has been slightly modified by the Kansas legislative task force. According to Raney Gilliland with the Legislative Resource Bureau, "The changes are pretty numerous, but the policies should be set." The bill is expected to allow utilities to recover stranded costs and transition costs for up to 12 years.

Kentucky: A legislative committee held numerous hearings and workshops on electricity restructuring in 1997 to prepare for a legislative proposal in 1998. The Public Service Commission has been involved in these activities, but has no formal investigation or docket.

Kentucky Utilities, one of the lowest cost suppliers in the nation, supports retail competition for all customers by Jan. 1, 1999. Kentucky Utilities favors federal guidelines for retail competition implementation, including a provision that would not allow utilities claiming stranded assets to compete for new retail customers until those stranded costs are removed.

Louisiana: Louisiana's Public Service Commission voted in December 1997 to present a procedural schedule for its January 1998 meeting to begin working out specifics for phasing in retail competition in the state. The potential benefits of competition were discussed in a series of meetings held in October 1997.

A legislative resolution was passed June 23, 1997, that established a 23-member committee to investigate retail competition issues and develop a legislative proposal for the 1998 session. This action follows the death of two restructuring bills that would have brought retail choice to rate payers beginning Jan. 1, 1999, and severely limited stranded cost recovery. The resolution also follows the Public Service Commission's denial of a proposed plan that permitted the recovery of all stranded costs over a seven-year period.

Maine: The issue of standard offer service was scheduled to be discussed at an inquiry opened by the Maine Public Utilities Commission in January 1998. The default service, similar to that in Massachusetts, is for customers who are either unable to take advantage of retail choice or are simply unwilling to participate. According to current legislation, standard offers must be in place by July 1, 1999, and be available until at least March 1, 2005. Also, the commission would like to address all stranded cost issues by July 1999, and all complexities concerning rate design by October 1999.

Maine was the third New England state to adopt retail choice legislation May 29, 1997. The measure brings retail access to about 640,000 end users in the state beginning March 2000. The law also requires utilities to sell off all generation assets before the start of competition and opens up metering and billing services to competition within two years of the start date. Utilities will be allowed to establish marketing affiliates, but are limited to supplying no more than 33 percent of the load in their transmission and distribution service area. Stranded costs are recoverable by the utilities, although not through an exit fee. Regulators will determine the amount of recovery allowed.

Maryland: On Dec. 30, 1997, the Maryland Public Service Commission voted to delay the opening of the state's retail electricity market by 15 months from dates set earlier in the month. The state's plan follows Pennsylvania's lead by opening the market to one-third of the state's load every year over a three-year period. Under a Dec. 3 order, a third of the state's end users would have begun receiving energy from alternative suppliers April 1, 1999, another third April 1, 2000, and the rest a year later. The new plan begins this cycle on July 1, 2000, and ends July 1, 2002. The commission pushed back implementation following a recent decision to call off a multibillion-dollar merger between Baltimore Electric and Gas, Baltimore, and Potomac Electric Power, Washington, D.C.

Massachusetts: On March 1, an electric industry restructuring bill will take effect that gives end users in the state an immediate 10 percent reduction on their bills. These savings will be guaranteed for seven years. Rates will be cut an additional 5 percent Sept. 1, 1999. Opponents of the bill argue that the plan is too soft on utilities, placing 100 percent of the stranded cost burden on Massachusetts rate payers Also, the reduction in savings from 15 percent in earlier drafts to 10 percent in the current version will cost end users nearly $200 million per year.

The state's Department of Public Utilities issued the final plan for restructuring the state's electricity industry Dec. 30, 1996. In the 400-page document, the department expressed its commitment to introduce full retail competition by Jan. 1, 1998. The plan followed a draft issued the previous May that called for the creation of an independent system operator and a separate power exchange, as well as for stranded cost recovery over a 10-year period through incentives for divestiture of generation and functional unbundling.

On July 25, 1996, Massachusetts became the fourth state to launch a retail wheeling plan when Massachusetts Electric, Westborough, Mass., began its program. Massachusetts Electric's industrial retail wheeling pilot was made available to 14 of the utility's industrial customers that belong to the Massachusetts High Technology Council, an industrial end user group that acts as an aggregator and buyer's agent.

Rather than having each member of the group shopping for their own power, the council selected Xenergy, Burlington, Mass., as the supplier for the whole group.

The utility expanded the program in January 1997 to include about 5,000 residential and commercial customers in four different geographic areas. Participants have saved about 15 percent.

Michigan: Detroit Edison, Detroit; Consumers Energy, Jackson, Mich.; and several business groups challenged a restructuring plan by the Michigan Public Service Commission that was to gradually open the state to competition over the next five years. These groups wanted a rehearing because of differences over issues such as the timetable for implementation. The request was denied in mid-January 1998.

The commission approved the plan in June 1997 to open the electric industry to competition by late 1997. The plan would have opened 2.5 percent of the total load of the major utilities in the state, Consumers Power and Detroit Edison, each year through 2001. The 2.5 percent allotment, approximately 150 MW for Consumers and 225 MW for Detroit Edison, would have been placed up for bid to all rate payer classes. Through the bidding process, an end user wishing to select an alternative energy supplier would be required to submit a bid detailing the amount the end user will pay in the form of a transition charge.

On Nov. 14, 1996, the Public Service Commission approved a plan created by Consumers Power to open at least 100 MW of load to a retail wheeling program. Under the program, participants were randomly selected through a lottery. Reportedly, end users representing more than 1,000 MW of load applied to participate. Potential suppliers to the program had to receive a certificate of public convenience and necessity from the commission. The commission used the transmission rate it developed in June 1995, but labeled part of it as stranded cost recovery so as not to cause a jurisdictional conflict with FERC. The commission previously ordered a five-year experimental retail wheeling program for 150 MW of industrial load for Consumers and Detroit Edison that began April 11, 1994.

Minnesota: The state's electricity restructuring task force had been expected to report to the legislature by January 1998 as required by previously passed legislation. This law was the lone survivor of a number of bills introduced in the 1997 session, many of which would have accelerated movement toward retail competition in the state. But the general lack of urgency among most end user groups is due primarily to Minnesota's low average electric rate of 5 cents per kWh.

Mississippi: On Nov. 3, 1997, the Mississippi Public Service Commission recommended a plan to implement retail wheeling in the state through a transition period from 2001 to 2004. Legislation would need to be passed in 1999 for this schedule to be met.

According to the bilateral contracts model the commission proposed, an independent system operator would play a reduced role as buyers and sellers would negotiate for energy and capacity in the marketplace. Stranded cost recovery will take place through an unavoidable wire charge from Jan. 1, 2001, to Dec. 31, 2004. The plan calls for rates to be unbundled by 2000, with competition beginning the following year. Noting that some assets will gain value while others lose value in a deregulated market, the commission suggested that stranded costs be dealt with on an individual basis. According to this plan, utilities would need to file their claims for stranded cost recovery by January 1999.

Missouri: The Missouri Public Service Commission opened a formal docket for investigating retail competition in the state to develop restructuring legislation. The commission formed a task force in March 1997 to conduct the inquiry, and a report is expected to be produced in February 1998. In addition, on March 28, 1997, the state legislature established a joint Senate-House committee to consider electric competition issues.

The commission approved a program offered by UtiliCorp's Missouri Public Service, Kansas City, Mo., that will allow commercial customers with at least 20 delivery points and a combined load of at least 2.5 MW to have access to competitive electricity suppliers. The program is scheduled to last two years and applies to 10 UtiliCorp end users.

As part of a settlement agreement among stakeholders in the Union Electric and Central Illinois Public Service merger, a 100 MW retail wheeling pilot program for all customer classes was approved by the commission. As part of the agreement, however, parties can still oppose the pilot once details are made available.

Montana: In accordance with the retail competition law passed in May 1997, Montana Power Co., Butte, Mont., presented its plan for competition. One hundred of the utility's largest customers, representing 40 percent of its total load, will be allowed to choose an energy supplier by July 1998. Another 10 percent of commercial and residential consumers will be eligible for choice by July 2000, with all consumers allowed supplier choice by July 1, 2002. Montana Power claims that to bring wide scale choice to end users, metering and billing systems will have to be rebuilt to handle the volume of end users seeking a new supplier of energy.

On May 2, 1997, Governor Marc Racicot signed the bill that made Montana the sixth state to enact retail competition legislation. Pilot programs, designed by each utility, will be required during the transition phase. Further provisions call for a two-year rate freeze at the beginning of the phase-in, followed by an additional energy component rate freeze for residential and commercial customers. Stranded costs will be handled through a transition charge based on filings made by the utilities one year before applicable customers gain direct access under the phase-in plan.

Nebraska: On Aug. 7, 1996, the state's single chamber legislature announced a three-year study of retail electricity competition. The purpose of the study, according to its sponsors, is to determine how the state can prepare for competition even though the entire state is fed by publicly owned utilities. The first phase of the study, already completed, reviewed deregulation issues within the state and around the country. The second phase of the study will result in specific public policy recommendations to be issued by the end of 1999.

Nevada: The state's rate payers eventually will be given retail choice according to a deregulation bill passed by the Nevada General Assembly and Senate July 6, 1997. The law sets the date for full competition in the state at Oct. 1, 2001, but this date could be advanced. Details of implementation, including stranded cost recovery, are to be determined by the Nevada Public Service Commission. Only a broad framework was established in the law. The commission was given the authority to license and revoke licenses of all alternative energy suppliers. The legislation was signed by the governor July 16, 1997, making Nevada the eighth state to pass a retail competition law.

Nevada enacted the country's first retail wheeling statute in 1994, but it only applied to new loads so as to attract industry to the state.

New Hampshire: The New Hampshire Public Utility Commission delayed the Jan. 1, 1998, start date for competition. Retail competition was pushed back to July 1, 1998. The delay was caused by a legal dispute between Northeast Utilities, the parent company of Public Service of New Hampshire, Manchester, N.H., and the commission.

Earlier, as part of this legal battle over stranded cost recovery by Northeast Utilities, the federal district court in Rhode Island issued a restraining order that prevented the commission from moving forward with any part of its restructuring plan that dealt with stranded costs recovery. The court later extended the restraining order for an indefinite period of time and included the enforceability of a 1989 agreement between the state and Northeast Utilities to the order.

The above controversy was sparked Feb. 28, 1997, when the commission released its final restructuring plan. The plan allowed for only 60 percent stranded cost recovery and eliminated any monopoly business in the competitive market. For utilities, that meant they had to divest all of their generation assets, or give up the opportunity to become a distribution service provider. Rate payers of Public Service were expected to see early savings of up to 19 percent as Northeast Utilities had not been guaranteed the recovery of all stranded costs. Northeast Utilities filed suit to block the plan March 1, 1997, claiming the restructuring guidelines went beyond utility guidelines by mandating that all industrial and commercial users move into the competitive market within six months of the market opening.

On May 22, 1996, New Hampshire was the first state to pass legislation that mandated the implementation of full retail competition for all customer classes by a certain date, in this case, mid-1998.

Three percent of the state's rate payers, or about 17,000 end users, began participating in the nation's first mandated retail wheeling pilot program in July 1996. The 3 percent was a randomly selected division of end users from all rate classes. The fight for the few larger commercial and industrial loads was so fierce that most participants were receiving price quotes below the market cost of power.

On another front, New Hampshire again made retail competition history in May 1996 when the state's Supreme Court upheld a Public Utilities Commission finding that there are no exclusive utility franchise territories in the state. Although full retail wheeling may make the point moot, the ruling was hailed as a victory for end users, especially in other states where the ruling could be used to set precedent. The case was based on the May 31, 1995, commission ruling that franchise areas in the state are not exclusive, which allowed Freedom Electric, Plymouth, N.H., to serve loads presently served by Public Service.

New Jersey: GPU Energy, Morristown, N.J., and Monroe township came to an agreement with Conectiv Energy, Wilmington, Del., Aug. 14, 1997, that placed New Jersey's only pilot program back on track. The plan, which stalled in May 1997 due to lack of supplier interest, is expected to bring savings of up to 5 percent to residential customers. Conectiv will be the electricity provider for about 10,000 residential and 850 commercial and industrial customers who agreed to let the municipality aggregate their load and explore the market for a more economical source of power.

On April 30, 1997, the Board of Public Utilities completed its plan to begin retail choice to the state beginning in July 2000. The board's plan requires a 5 percent to 10 percent rate reduction for all customer classes starting October 1998 when 10 percent of end users will be able to choose an alternative energy supplier. Retail choice will become available to all end users by June 2000, by which time they can expect to see rate decreases of up to 15 percent as utility assets are paid off and proposed state energy tax cuts are made. Stranded costs will be evaluated and awarded on a utility-by-utility basis, provided the electric companies can demonstrate they have taken significant measures to control and reduce stranded investments.

In response to the board's call for full retail choice by mid-2000, Public Service Electric & Gas (PSE&G), Newark, N.J., offered a more direct approach to competition. PSE&G proposed that all of its customers be allowed to participate beginning Jan. 1, 1999. Customers would submit their choice of energy supplier by October 1998. The proposed plan includes a seven-year rate cap and a 5 percent to 10 percent rate cut for all consumers. PSE&G said it plans to secure $2.5 billion of its alleged $5.5 billion in stranded costs to provide the rate cut.

Governor Christine Whitman signed a bill into law July 15, 1997, that eliminated the gross receipts and franchise tax placed on utilities, replacing it with a straight sales tax to be levied against all electricity suppliers. The tax reform is intended to simplify tax collection from alternative suppliers from outside the state as well as provide a tax cut to end users.

On Jan. 16, 1997, the New Jersey legislature proposed it's plans for restructuring. New Jersey's "Energy Master Plan" combines the allowance of bilateral contracts and the establishment of a power pool coordinated by an independent system operator. The plan calls for 5 percent of utility load to be made available to the open market beginning October 1998, and gradually builds to 100 percent by April 2001.

New Mexico: On Sept. 9, 1997, the city of Albuquerque submitted a proposal to the New Mexico Public Utilities Commission to implement a limited pilot program. Three end users, already selected, would participate. These end users include a wood processor, a 96-unit residential community, and the city's water department. The city wanted the program to begin in January 1998 and last for one year with the intent of serving as a model for the rest of the state.

After a breakdown in deregulation discussions in July, New Mexico prepared a report that detailed points of compromise and those issues that could not be agreed upon among the parties in the deregulation debate. Utilities preferred a later start to competition, suggesting that an earlier date would infringe on their ability to recover a sufficient amount of stranded costs. They suggested start dates ranging from 2001 to 2004; groups representing environmental concerns, industrial consumers, and competing suppliers are pushing for a start as early as 1999. The goal of this stockholder group is to develop a legislative package for the next session. A series of retail wheeling bills were introduced in the 1995 New Mexico legislature: all were killed early due to utility opposition.

On March 27, 1997, Texas-New Mexico Power, Fort Worth, Texas, had a retail competition plan approved by the Public Utilities Commission. The "Community Choice" plan, originally filed in June 1996, was scheduled to implement retail wheeling over a three year period. Beginning April 1, 1997, the utility froze rates until 2000, during which time it will recover its stranded costs. In 2000, all end users will have access to competitive suppliers, and smaller users will be allowed to aggregate to increase buying power. Further, a limited pilot allows 1 percent of the utility's customers to choose an alternative supplier during the three-year transition. This plan was the first utility restructuring plan filed with the commission. This plan also was filed in Texas and later withdrawn, but the utility is still pursuing the plan in New Mexico.

The commission approved a negotiated agreement for large industrial end users of the Plains Electric Generation and Transmission Cooperative, Lubbock, Texas, that allows 1 percent of existing load and all new load to purchase electricity on the open market. Although the original agreement only applied to certain industrial rate payers, the Public Utilities Commission amended the program to include a portion of all customers of the 13 distribution cooperatives. The implementation date of the program is unknown.

New York: The New York Power Pool currently has a proposal
before the New York Public Service Commission asking for a
competitive transition charge to recover the shortfall between
its member's market revenues and their generators' total going forward costs. The Independent Power Producers of New
York Inc. argue that if this practice is allowed, the utilities
will be empowered to underbid their capacity and energy in
a free market without suffering financial loss.

In an effort to bring competition to the metering side of
the electric power industry, the commission requested that
some large industrial and commercial end users be given the
opportunity to purchase their electric meters from the state's
investor-owned utilities. While the utility will still maintain
control of the meter, the commission feels that this is a good
way to test interest in meter ownership.

Although there was significant energy-related legislative
activity in 1997, no general restructuring bill passed before
the end of the session. The assembly did pass a bill, called
"Competition Plus," that calls for full retail customer choice
by September 2000. The bill also mandates the creation of an
independent system operator and the functional unbundling
of the state's utilities. The stranded cost issue is not fully addressed, and there is no guarantee of full recovery for the
utilities included in the bill. In 1997, the state Senate passed a
bill to phase out the utility gross receipts tax, but no restructuring bill was completed.

Despite limited legislative progress, the state's Public
Service Commission continues to move forward with its own
restructuring initiatives, assuming no legislation is required
to give them this authority. On May 16, 1996, the commission
issued a plan to introduce retail competition to the state, and
it surprised a few stakeholders by suggesting that retail
wheeling be offered to all customers in all classes by 1998.
The plan required all state utilities to file restructuring procedures that detail how retail wheeling will be implemented in

this time frame; show how they will restructure their companies; provide a rate plan; and address certain social issues. The restructuring proposals originally were due October 1, 1996, but the deadline was extended to March 10, 1997.

The year following this order introduced a steady stream of negotiations between the commission and state utilities.

The most hotly contested of these plans, Consolidated Edison's, was settled in September 1997, so that retail access will be available to all end users beginning June 1, 1998, and phased in through 2001. Along with guaranteed rate decreases of 10 percent for small end users, Consolidated Edison agreed to divest 50 percent of its generation assets in New York City. The settlement with New York State Electric and Gas, Binghamton, N.Y., will bring retail competition to end users by August 1999. Industrial end users will receive a rate reduction of 5 percent per year spanning a five-year period, resulting in savings of 22.7 percent overall. Smaller users will avoid a planned rate hike and have their rates frozen. Rochester Gas and Electric's proposal will phase in choice over a five-year period beginning in 1998. Commercial and industrial end users will receive a 10 percent rate reduction.

Orange and Rockland will open their market in May 1998, with choice for capacity following in May 1999. Industrial users will see rate reductions of up to 12 percent under the agreement, while smaller users will only see savings of about 2 percent. Central Hudson, which already has rates below the commission's target rates, will freeze its rates for all end users through July 2001. The utility will be awarded full stranded cost recovery, but will have to install real-time metering for all end users.

Niagara Mohawk, Syracuse, announced its restructuring proposal Oct. 5, 1995. Its plan calls for splitting the utility into separate companies for generation, transmission, distribution, and marketing. It also details voluntary poolco and bilateral contracts that would allow direct retail access within three years.

The New York Power Authority approved agreements Jan. 29, 1997, for the elimination of the New York Power Pool and the establishment of a nonprofit independent system operator for the state. The ISO will be responsible for balancing load with power generation, economic dispatch of power through a newly established power exchange, and the creation of some pricing structure. The ISO will have the oversight of 28 representatives from utilities, power marketers, environmentalists, and consumer groups. The plan also establishes the New York State Reliability Council to set standards in accordance with national, regional, and state regulations. Local utilities will be guaranteed all required revenue associated with the operation of transmission facilities through a transmission service charge to customers. The plan has been submitted to FERC. The Power Authority hopes to complete the implementation by the middle of 1998.

North Carolina: In an attempt to get the legislative process for electric industry restructuring moving, the North Carolina General Assembly approved a bill that established a commission to study the issue. The committee is composed of 23 stakeholders representing the state's five largest electric utilities, consumer groups, and state senators and representatives. The commission, which had to issue an interim report in January 1998 and will issue a final report in January 1999, is investigating all aspects of restructuring from environmental impacts to stranded cost resolution.

After passage of the commission bill, a separate plan was introduced to bring retail choice to the state starting in 1998. The "Customer Choice in Electricity Act" would give residential consumers priority in allowing them to choose their electricity supplier beginning Oct. 1, 1998. Commercial and industrial end users would be allowed access to alternate suppliers in January and July of 1999, respectively. Investor-owned utilities would be granted 50 percent stranded cost recovery over a period of five years. Municipal and cooperative energy suppliers, on the other hand, could choose among three options: 100 percent cost recovery if they remove themselves from the electricity market entirely; 50 percent recovery if they choose to allow competitive electricity supply into their service territory; or close their system to competition.

North Dakota: Two bills have been introduced to the North Dakota legislature, one from each house, that mandate the study of other retail competition initiatives. This will allow restructuring legislation to be considered in the 1999 session. On Sept. 13, 1996, the North Dakota Public Service Commission issued a second order relating to electricity industry restructuring—the first opened the investigation—that provides for continuing investigations into retail competition.

Ohio: On Jan. 6, 1997, the Ohio General Assembly announced the creation of a joint legislative committee to study retail competition and restructuring issues. The committee likely reviewed previous legislative initiatives with an eye toward developing a new proposal. The committee was expected to release its report Oct. 1, 1997.

The Ohio Supreme Court produced a ruling that could have a great impact on the issue of restructuring in Ohio, possibly delaying it until 1999. The court ruled that funding for the state's educational system is unconstitutional, affecting restructuring because a majority of school funding comes from utility taxes. The Joint Select Committee appointed to study restructuring had planned talks regarding utility tax redistribution in the near future, but it is unclear how the ruling will modify the overall re-regulation process.

On Dec. 24, 1996, the Public Utility Commission of Ohio issued guidelines on a conjunctive billing program that allows smaller end users or end users with multiple facilities to aggregate their load and take advantage of load diversity and greater purchasing power.

This proposal is being implemented through a two-year pilot program that requires each aggregated group to negotiate the rate that would apply to the total load. Billing and metering can be handled by a third party aggregator, but the supplier of the energy is still the native utility.

The Conjunctive Electric Services tariff is mandatory for all Ohio utilities, and contrary to earlier designs that called for revenue neutrality, the approved plan allows customers to save money through aggregation, as they would in a competitive market.

Oklahoma: State Sen. Kevin Easley introduced a bill in late January 1998 to speed up completion of studies related to electricity competition as required by a 1997 law he supported. Easley proposed the move because a related study, which dealt with the possibility of creating an independent system operator (ISO), was completed on time. State Rep. Chris Hastings introduced a bill that would push the goal of starting electricity competition in the state from July 1, 2002, to July 1, 2000.

The Oklahoma Electric Restructuring Act of 1997 was the first legislation of its kind in the region, which already is characterized by below average rates. It set a deadline for competition at July 1, 2002, and froze rates until competition began. The stranded cost recovery period was set at seven years and utilities are prohibited from energy rate increases due to cost recovery.

Oregon: On Oct. 21, 1997, the Oregon Public Utility Commission approved a pilot program by Portland General Electric, Portland, that will allow about 50,000 end users in its service territory a choice of energy supplier. The commission, however, suspended the program until remaining issues can be resolved. When implemented, participants will account for about 15 percent of the utility's load.

Portland General Electric submitted a restructuring proposal on Dec. 1, 1997, that would transform the company to a regulated electricity transmission and distribution company. The restructuring includes selling all of its generating assets through a competitive bid process and enacting a 3 percent "system benefit charge" to end users to support public purposes under regulatory mandates, such as energy conservation measures and low-income weatherization. The plan would give end users power price decreases of about 10 percent, which would be in addition to decreases of 7 percent provided in December 1996.

PacifiCorp, Portland, Ore., announced a competition plan Oct. 7, 1997, that would allow all end users with a demand greater than 5 megawatts (MW) to choose a power supplier for up to 50 percent of their load. The company's proposal also would allow 30,000 end users of all classes in Klamath Falls, Ore., and all schools and universities in the state to receive their energy needs from alternate suppliers. The plan was to be filed at the end of 1997.

Despite the teaming of two major proponents of electric industry deregulation, legislation drafted by The Fair and Clean Energy Coalition and the Oregon Energy Coalition failed to pass through the Oregon legislature. This deregulation bill would have brought retail choice to large industrial companies beginning Dec. 1, 1999, and to all consumers by Oct. 1, 2001. The measure also addressed environmental concerns through the inclusion of a 3 percent dedication of total utility revenues to energy conservation and renewable resource generation research programs. Oregon was part of a four-state steering committee that finalized an electricity restructuring plan Dec. 5, 1996, that aims to keep low-cost power in the region.

Pennsylvania: In mid-December, the Pennsylvania Public Utility Commission offered its own plan to restructure electricity sales within the territory of Philadelphia-based Peco Energy, rejecting competing proposals by Peco and Enron, Houston (see 1998 January EUN). Under the PUC plan, electricity bills are expected to be reduced by about 15 percent. Two-thirds of Peco's end users will have retail access by Jan. 2, 1999, while all end users will have access by Jan. 2, 2000. Peco will recover about $4.9 billion in stranded costs from Jan. 1, 1999, to June 30, 2007, by assessing all end users a competitive transition charge. Peco has appealed the plan.

On Dec. 3, 1996, Pennsylvania became the fourth state to pass the legislation necessary to implement retail electricity competition. The bill, called the Electricity Generation Customer Choice and Competition Act, easily passed both houses after several weeks of consensus building among the state's stakeholders. Some key provisions of the law state that competition will be phased in over three years from Jan. 1, 1999, to Jan. 1, 2001, by adding one-third of the state's load each year; the commission could mandate pilot programs, beginning with 5 percent of each utility's load in 1997; each utility was required to submit a restructuring plan to the commission, which had to make stranded cost decisions by Sept. 30, 1997.

Rhode Island: On July 1, 1997, retail choice became available for manufacturing firms with an average annual demand of 1,500 kilowatts (kW) or more (see September 1997 EUN).

As of EUN press time, the Rhode Island House of Representatives was reviewing a bill that will answer many lingering questions surrounding the introduction of competition to the state. The bill, sources say, does not back off the state's rapid phase-in schedule or increase stranded cost recovery detailed in the original legislation, as some utilities had hoped. The main focus of the bill is to provide exemptions for some of the smaller, non-competitive utilities, and gives regulators flexibility to grant alternative stranded cost recovery methods to utilities if the method lowers charges to consumers.

On Aug. 9, 1996, Governor Lincoln Almond signed legislation that required full retail competition for all end users and allowed some end users direct access by mid-1997. Both houses of the legislature passed the law, called the Utility Restructuring Act of 1996, after some major revisions were made from a bill negotiated by Narragansett Electric Co., Providence, which serves 80 percent of the load in the state. According to the plan, 10 percent of the state's load was to have access to competitive markets by July 1, 1997—selected on a first-come, first-served basis—20 percent by Jan. 1, 1998, and all remaining load by July 1, 1998.

South Carolina: In late January of this year, the state's Public Service Commission unanimously rejected a request by Electric Lite, Greenville, S.C., to implement a customer choice program that would have allowed Electric Lite and other alternative energy suppliers to buy wholesale from utilities and resell it to end users. The suppliers would have paid the utilities a fee set by the commission.

South Carolina's legislature was not able to reach a consensus on an electricity restructuring bill before the 1997 session came to a close. At the close of the session June 5, 1997, the House was considering a bill that would have brought retail choice to all customers in the state by 1999, starting with residential customers Jan. 1, 1998.

The Public Service Commission voted to begin accepting proposals for deregulation May 15, 1997. In response to the Commission's request, the state's major public utilities and consumer groups entered their suggestions in late June last year. Duke Energy took the most proactive utility stance on the issue, outlining the establishment of an independent system operator, a formal consumer education process through a third party, and customer choice of energy providers. Duke's plan did not suggest a date for the process to begin, but a slow evolution seems to be the desired outcome given the comments submitted by the state's other utilities, Carolina Power & Light and South Carolina Electric and Gas.

South Carolina joined the ranks of states with formal restructuring bills Feb. 6, 1997. The legislation introduced by Rep. Doug Smith called for an aggressive schedule bringing retail choice to commercial and industrial end users by July 1, 1998, and Jan. 1, 1999, respectively.

South Dakota: South Dakota law allows new customers with usage of two or more MW to petition the Public Utilities Commission to select their electric suppliers. Smaller users must contract with their assigned utility.

Tennessee: On June 19, 1997, the governor signed legislation mandating a study into retail electricity competition issues. A major issue in Tennessee is the impact of deregulation on the Tennessee Valley Authority and the subsequent impact on local ratepayers. The committee was scheduled to issue a report and proposed legislation by February 1998.

Texas: On Dec. 22, 1997, Texas-New Mexico Power Co., Fort Worth, Texas, reached an agreement with the state's Public Utility Commission on a proposal for a five-year transition to competition. Prices were reduced by 1 percent for commercial end users beginning Jan. 1, 1998. Equal reductions will occur January 2000 and 2001, for a total reduction of 3 percent for commercial end users. Originally filed July 31, 1997, the plan is supported by more than 80 percent of the communities served by the utility through ordinances and resolutions.

Despite attempts by Governor George Bush to push electricity restructuring legislation, Texas will have to wait until at least the next biennial legislative session in 1999 to pass a meaningful deregulation bill. Bush introduced a last-ditch compromise bill to the House last May, but it did not pass due to resistance from stakeholders who opposed the 100 percent stranded cost recovery by utilities and delayed start date of 2002 established by the bill.

The restructuring plan unveiled by state Sen. Jerry Patterson and state Rep. Mark Stiles in February last year was stalled in committee hearings before Lt. Governor Bob Bullock placed the measure with another committee. The bill called for an immediate 15 percent rate decrease to end users by September 1997. Schools, colleges, and universities would choose their suppliers by August 1998, followed by light commercial end users in Jan. 1999. Large industrial customers would be last on the list, able to choose Jan. 1, 2000.

Texas became the first state to approve an independent system operator, which was to be operational by 1997 in a plan filed with the Public Utilities Commission, as part of Texas' wholesale market restructuring. Besides implementing non-discriminatory transmission access, the ISO would implement a transmission information system as required in FERC Order 889.

Utah: A legislative task force was created by a law passed in March 1997 to study electricity competition and develop an implementation plan. An initial report was due November 1997 with a final report and suggested legislation due by November 1998. The task force also was charged with addressing stranded cost issues.

On Jan. 24, 1996, the Utah Public Service Commission issued a notice of inquiry on electric industry restructuring, thus opening a formal investigation into retail competition.

Vermont: After months of debate in the state Senate, the Vermont House of Representatives rejected an intensive electricity restructuring bill in the 1997 session, which ended the first week of May. Many of the utilities in the state were disappointed by the House's refusal to pass the Senate bill, which underwent months of revision and debate before passing. The House plans to create its own restructuring bill early in the next session.

The Senate's bill would have brought full retail competition to all end users by the middle of 1998. The bill called for a 50-50 split of stranded costs between utility shareholders and consumers, but utilities claimed that would bankrupt them. In a revised version of the bill, regulators would evaluate individual utilities' financial condition to determine how stranded costs could be "divided evenly" between rate payers and shareholders. The bill also included language mandating that imprudent expenditures would be ineligible for cost recovery.

The state's Public Service Board released its final restructuring report Dec. 30, 1996. Under pressure from end user groups and a few utilities, the board accelerated the date the state electricity market would open to direct access for all end users. The draft restructuring report, issued in October 1996, suggested phasing in competition through 2000, but the final report called for full access for all end users by the end of 1998. The board, however, elected to slow the process to remedy stranded cost concerns, setting a date of December 2001 before a final decision is made.

Virginia: Two competing bills for restructuring the electricity industry were introduced to the state legislature in late January 1998 by Sen. Jackson Reasor Jr. and Rep. Kenneth Plum. Reasor's bill would lead to full competition in 2004. Plum's bill gave no exact date for retail competition, but said that wholesale power competition would begin Jan. 1, 2004. The pair have since consulted to form a compromise bill.

The staff of the Virginia State Corporation Commission, at the request of the state General Assembly, recommended a cautious, two-phase, five-year plan to study the feasibility of implementing retail access. During phase one, each utility's rate structure would be thoroughly analyzed and small pilot programs would be started. This phase would last from 1998 to 2001 and would involve the formation of an independent system operator and a regional power exchange. The state would determine how and when to implement full retail choice during phase two, lasting from 2000 to 2002.

Washington: A bill introduced to the legislature in January 1998 by state Sen. Lisa Brown and others could help preserve the low cost and high reliability of power in Washington. The bill requires utilities and others to make extensive disclosures of rates, credit policies, and dispute resolution procedures. Utilities also would have to identify the types of resources used to generate power, and the air emissions they produce.

On Oct. 1, 1997, Washington Water Power, Spokane, Wash., unveiled plans for a pilot program that would effectively replace the More Options for Power Service pilot that has been giving access to commercial end users in two cities since July 1, 1997. Local Access Market Pricing, which could begin in early 1998, would give end users the option of selecting their electric service from a menu that would list prices as a function of energy type and terms of service. Those who are willing to purchase metering devices, costing between $150 and $800, would be given direct access to energy markets and a choice of electricity providers. The state's Public Utility Commission approved a separate retail wheeling pilot program in September 1996 that gave Washington Water Power's largest end users the opportunity to choose competing suppliers for up to one-third of their total load.

The pilot coordinated by Puget Sound Energy, Bellevue, Wash., which began on Nov. 1, 1997, allows up to 85,000 end users to select an alternative energy supplier. End users are allowed to select a bill from their energy supplier and Puget

Sound for transmission and distribution, or receive a consolidated statement from the utility. To increase interest and participation in the pilot, Puget Sound offers a delivery discount to participants ranging from 1.5 percent to 9 percent.

Washington's legislative attempt to bring a comprehensive restructuring bill to a final vote in 1997 proved to be an insurmountable task due to heavy utility opposition. Legislation that would have brought full competition to the state beginning July 1, 1999, and allowed stranded cost recovery until June 30, 2004, received considerable changes before passing the state Senate March 18, 1997. The drastically weakened bill required utilities to unbundle billing statements showing transmission, generation, and distribution as separate components beginning Oct. 1, 1998.

Companion retail competition bills were introduced to both legislative houses in the first week of February 1997, calling for a five year phase-in period starting mid-1999. Called the Electric Consumers Act, the bills call for a 50-50 split between shareholders and customers for stranded cost recovery over the five year transition period.

Washington was part of a four-state steering committee that finalized an electricity restructuring plan Dec. 5, 1996, that aims to keep low-cost power in the region.

West Virginia: The West Virginia Public Service Commission initiated a formal investigation into retail competition by asking for input from various stakeholders on key issues, such as stranded costs, customer access, and overall economic benefits. Hearings began in April 1997, and a report was due by Dec. 22, 1997, to complete model legislation.

Wisconsin: On Aug. 13, 1997, the Wisconsin Public Service Commission issued a draft order to revise the 32-step plan issued in 1995 to bring retail electricity competition to the state. In the order, the commission addressed reliability issues due to threats of capacity shortages during the summer of 1997. With respect to competition, the draft order suggested moving the full access date to February 2000, although this is the earliest possible date since the state legislature will not address restructuring legislation until the 1999 session.

While Wisconsin is not one of the first states with retail wheeling, it seems to be leading the pack in the development of an independent system operator. A coalition of stakeholders, including the state's industrial end user group, developed an ISO structuring proposal that has become a model for other state groups. Meanwhile, the commission formed a task force to develop their own proposal, but the group consists only of three utilities and a rate payer advocacy group.

On Dec. 19, 1995, the Public Service Commission approved the original plan to introduce retail competition by the year 2001, if certain market conditions are set. The plan called for a 32-step process to phase in competition over five years. The plan covered the resolution of some key issues, such as open access transmission, creation of an independent system operator, the risk of exporting low-cost power generated in Wisconsin, affiliate dealing, and social programs.

Wyoming: As a result of a series of initial meetings, the Wyoming Public Service Commission issued a draft white paper on various electricity industry restructuring issues. In this investigation the commission formed six groups, each of which tackled different issues, such as legal issues, transition costs, and pilot programs. The commission found that it does not have the legal authority to mandate restructuring or competition, and legislative changes will be necessary. The commission recommended that utilities should be able to recover stranded costs and should develop a pilot program.

Chapter 3

Lessons Learned from Deregulation in Scandinavia & England

The Association of Energy Engineers (AEE) conducted a study mission to northern European countries during the time frame of July 1-18, 1997. The itinerary included England, Holland, Germany, Denmark, Sweden, and Norway. AEE delegates were received by various U.S. Embassy, local country government energy and environmental officials, utility executives, and energy business personnel. Meetings were held in London, Amsterdam, Copenhagen, Stockholm, and Oslo. Subject matters discussed focused upon electric deregulation, cogeneration, district cooling, energy rates, and current energy and environmental policies.

LONDON

London meetings were conducted with various groups including the U.S. Embassy, The Electricity Association, PowerGen, and the European Bank of Reconstruction. Discussions focused upon the deregulation of the electric power industry, the current

Presented at the 1998 Competitive Power Congress by Kenneth J. Kogut, P.E., CEM windfall tax, and investment in energy efficiency for various building sectors and groups.

As was determined, the United Kingdom has deregulated, or privatized, its electric distribution systems since 1990. This represents one of the oldest deregulated electric systems in the world today.

Under the previous structure, the nationalized United Kingdom electric system was dominated by one large generating and transmission company, the Central Electricity Generating Board (CEGB). The CEGB sold electricity in bulk to 12 area distributing boards, each of which served a closed supply area or franchise. Under the new restructuring of the electric utility industry, the old CEGB was split into four parts as of April, 1990. The power stations were divided between two large fossil-fired generators, National Power and PowerGen, and a nuclear generator, Nuclear Electric. National Power and PowerGen joined the private sector, while Nuclear Electric remained in public ownership until 1996.

At the time of privatization, National Power became responsible for 30,000 MW of capacity, and PowerGen the remaining 18,000 MW of fossil fuel-fired plant. Some 8,400 MW of nuclear capacity previously owned by the CEGB was allocated to Nuclear Electric, and 2,100 MW of hydroelectric pumped storage capacity to the new transmission operator, NGC. The pumped storage business, First Hydro, was transferred from NGC in 1995 and is now owned by Mission Energy of the United States.

All this has led to an increasingly competitive market for base load generation. Forty-six (46) generation licenses have been issued in England and Wales since privatization, and there are already at least twenty-two (22) independent generators selling electricity into the Pool.

There are several key features to the new system. First, power is traded through an open commodity market, the Pool. Second, the generators no longer have any obligation to supply, or any assured market. They have to compete for their share of an increasingly competitive market. All the major generating companies are required to sell the electricity they produce into an open commodity market known as the Pool.

Essentially, each generating unit has to declare by 10 am each day its availability to the market, together with the price at which it is prepared to generate, for each and every half hour of the following day. The units are then called to generate by the NGC in ascending order of price.

The most expensive unit used established the system marginal energy price which all others receive for that half hour. There is an additional separate pricing mechanism designed to provide an incentive for the provision of generating capacity. About 90 percent of the electricity sold by the major generating companies is covered by contracts, both with the RECs and with individual large customers. Only around 10 percent of electricity sold is paid for at Pool prices.

The market has had an effect on all those who sell into it. As electricity cannot be differentiated by source of quality, the challenge is to be the least cost producer. All the generating companies have implemented a range of measures to reduce costs and have diversified fuel sources and the range of fuels used. This means burning gas in new, more efficient CCGT plant, and securing supplies of gas by contracting for independent gas production or joining with others in exploration. Companies are seeking to buy coal at world prices, whether from the UK or overseas, taking advantage of low sulphur coal. These measures have already led to major reductions in fuel costs.

It has been claimed, particularly by those opposed to a more open electricity market, that most customers have suffered substantial price increases since privatization. During the early years of privatization, prices rose during the period of relatively high inflation from 1989 to 1992.

In 1996, domestic electricity prices were 11 percent lower in real terms than at privatization. Current cost for electric for the industrial customers is approximately $0.08/kWh (USD), and approximately $0.10/kWh (USD) for residential without the VAT included. When VAT is included, UK domestic prices are still among the lowest in the European Union—32 percent lower than in Germany, 38 percent lower than in Belgium, 31 percent lower than in France, and 27 percent lower than in Spain. UK prices for moderately large industrial customers with a demand of 2.5 MW, based on published tariffs, are in the middle of the price range. However, UK industrial customers who buy electricity under competitive contracts enjoy even lower prices which are the fifth cheapest among the 15 Member States of the European Union. Additional price reductions can be expected as the fossil fuel levy is further reduced.

While much has already been achieved since privatization, many challenges remain. On the generation side, efforts must be made to ensure that competition continues to drive down costs and that these savings feed through into further price reductions for customers. In the regulated parts of the business, electricity companies will have to meet more demanding targets set by the Regulator, which are imposing real price reductions from year to year.

The mechanisms of the electricity marketplace will also evolve. At present, the Pool in England operates as a producer's market and the demand-side is in the early stages of development. New arrangements are presently under trial to test the scope for demand-side bidding in the Pool, whereby customers would be given incentives to reduce their consumption at certain times. There is already evidence that this is exerting a downward pressure on Pool prices. Progress can also be expected in the treatment of transmission costs, in particular those costs associated with transmission constraints. Over time, a longer term forward market in electricity may also develop.

Finally, the issue of energy efficiency will also take on greater importance. Electricity companies have already taken on a number of initiatives in the area of demand-side management and the question of incentives to promote greater efficiency is being examined both by the industry and the Regulator.

On July 2, 1997, the Chancellor announced the introduction

of the proposed windfall tax on the excess profits of the privatized utilities. The one-off tax will apply to companies privatized by flotation and regulated by statute. The tax will be charged at a rate of 23 percent on the difference between company value, calculated by reference to profits over a period of up to four years following privatization, and the value placed on the company at the time of flotation. The expected yield is around 5.2 billion Pounds.

The tax will apply to companies privatized by flotation and regulated (or with a subsidiary regulated) by relevant privatizing statutes. These are the Telecommunications Act 1984, the Airports Act 1986, the Gas Act 1986, the Water Act 1989, the Electricity Act 1989 (and the Electricity (Northern Ireland) Order 1992) and the Railways Act 1993. The tax will be charged on the floated companies.

The companies falling within the scope of the windfall tax are BAA, British Energy, British Gas (now BG plc and Centrica), British Telecom, National Power, Northern Ireland Electricity, PowerGen, Scottish Hydro, Scottish Power and Railtrack, the regional electricity companies and the privatized water and sewerage companies (including such companies now forming part of Hyder, United Utilities and Scottish Power).

The Electric Association, equivalent to the Edison Electric Institute, is not in favor of the Windfall Tax. In the end, it is believed that it will adversely affect deregulation, overall, and the resultant savings derived from a deregulated electric industry.

Meetings were conducted with the European Bank of Reconstruction (EBRD) and included discussions focusing upon the implementation and financing of energy efficiency measures in various sectors of the world economy.

The EBRD is a multinational institution set up with the specific aim of assisting the countries of central and eastern Europe and the CIS to develop into market-oriented economies. Its shareholders include countries from both this region and the rest of the world, plus the European Community and the European Investment Bank.

Specifically, the EBRD seeks to promote the development of the private sector within these economies through its investment operations and through the mobilization of foreign and domestic capital. The EBRD's main advantages, compared with private commercial banks, lies in its willingness and ability to bear risk, as a result of its shareholder base. This allows the Bank to act at the frontier of commercial possibilities and to be an effective "demonstrator." It also shares the project risk by acting with other private sector entities, such as commercial banks and investment funds, as well as multilateral lenders and national export credit agencies. The EBRD assists companies that have difficulty in securing financing: as such, it complements the efforts of other lenders.

Typically, the EBRD funds up to 35 percent of the total project cost for a greenfield project or 35 percent of the long-term capitalization of an established company.

Currently, the EBRD is developing framework agreements for the start-up of energy service companies (ESCO's) in Central Europe. In addition, they are investing in energy efficiency producing projects such as improved district heating, networks, and several projects aimed at reducing the energy consumption in large industrial companies. Cities that the EBRD is working with include Kiev and other cities in the Ukraine, several cities in Russia, Sofia and Mirisk. AEE has laid some groundwork in further developing an ongoing business relationship with the EBRD and AEE's expanding chapters in eastern and central Europe.

AMSTERDAM

AEE delegates met with utility executives from NV UNA. Discussions focused upon the controlled and limited deregulation market currently existing in Holland and Dutch generating companies. UNA currently has a generating capacity of 3.6 gigawatts (GW), and ranks third in the Holland utility electric utility industry, with EPZ rated at 4.6 GW and EPON rated at 4.3 GW ranked above UNA. Discussions focused around the possible merging of the three utilities pushing them to an overall ranking of eleventh within the European electricity companies.

Currently, there are four (4) Dutch generating companies: EPZ, EPON, UNA, and EZH. Prior to the current controlled form of deregulation, there existed fifty (50) distributing companies. Subsequent to deregulation, there currently exists twenty (20) distributing companies.

The present electric/heat generating plant fuel mix of UNA consists of 62 percent gas, 20 percent coal, 17 percent Blast Furnace Gas (BFG), and 0.41 percent oil. This is compared to the Netherlands generating fuel mix of 42 percent coal, 46 percent gas, 4.2 percent BFG, 0.3 percent oil, and 7.9 percent nuclear. The Netherlands electric energy production is estimated to be 40 GWh/year. Coal imports for electric producing plants are imported from Australia, Columbia, and South Africa. Domestic coal mines have been shut down.

Peak electric loads for the Dutch generating system is 11,700 MW, with an estimated system capacity of 10,000 MW. Typically 86 percent of system peaks are generated, with 14 percent being imported.

The Netherlands are currently working under an order of Economic Affairs titled Streamlining '96. Some major points of this directive include:

- Keep transmission and distribution separate, an "arms length" from generation.

- Merger of the four (4) producers—Create a "Great Production Company."

- Provide special status for sustainable energy—ex: solar should be protected.

- A new government body would be created governing the gas and electric business. The new body would control free traffic of energy and protect "captive" users.

As a result of current energy operations, downtime in the Netherlands is approximately six (6) minutes per year. This is compared to thirty (30) minutes per year for the average in Europe.

Based on the present electric energy rates, approximately 60 percent of cost per kWh goes to the price of fuel. The general price for electricity for the industrial sector in the Netherlands is $0.065/kWh (USD), of which, approximately less than $0.01/kWh represents the systems grid (380 kVA) cost which is needed to be supported. Overall the Netherlands maintain a 65-70 percent systems load factor.

Up until recently, electricity in the Netherlands has been free of tax. Households currently have a value added tax (VAT) on energy. Residential rates with VAT added are at approximately $0.15/kWh (USD), and approximately $0.11/kWh without the VAT.

Privatization is not in the near picture for the Netherlands. Some competition exists, but rates are relatively low compared to other European countries.

COPENHAGEN

The Ministry of Environment and Energy held court for AEE delegates in Copenhagen, Denmark. Discussions focused upon development energy policies, developed by The Danish Energy Agency, as related to the country of Denmark. The agencies of energy and environment were combined since 1994.

The Danish Energy Agency, an Agency under the Ministry of Environment and Energy was established in 1976. The Agency focuses on the production, supply and consumption of energy and ensures, on behalf of the State, the responsible development of energy in Denmark from the perspectives of society, the environment and security of supply.

The Agency drafts and administers Danish energy legislation

and implements analyses and assessments of development in the energy field. One of its important tasks is to launch initiatives to translate Government energy policy into practical initiatives. In April 1996, the Government presented its action plan for energy, *Energy 21*, the aim of which is to contribute Denmark maintaining and developing its pioneering role in the achievement of sustainment of sustainable global development. In 1996, 8 percent of the total Danish energy consumption was covered by renewable energy sources. According to *Energy 21*, this share shall increase to 30 percent in 2025.

Energy 21 maintains the Government's objective of reducing CO_2 emission in Denmark by 20 percent by the year 2005 in relation to 1988 levels.

The Danish Energy Agency is responsible for the overall planning of power, heat and natural gas supplies in Denmark. In addition to electricity, most consumers today have access to energy supply networks consisting of district heating or natural gas. In areas not supplied by district heating or natural gas, various subsidy schemes provide incentives to consumers to switch environmentally friendly forms of heating.

Energy supply networks are in the process of being made more efficient by increasing co-production of heat and power at combined heat and power (CHP) plants. CHP supply from large-scale plants is being further developed and local district heating plants converted to CHP based on natural gas, waste and biomass. New, local district heating systems are also being established and industries are being encourage to establish their own, local CHP Plants.

Denmark is also a leader with regard to energy savings. The Agency works to encourage the general public, trade and industry and public institutions to make even more efficient use of energy in order to achieve savings. Initiatives targeting private consumers include energy conservation campaigns and various subsidy schemes. Labeling of electrical appliances is being introduced to encourage consumers to purchase low-energy appliances.

The Energy Agency also administers a scheme launched in 1997 to energy-label all buildings so that buyers know the energy condition of a building when contemplating a purchase. Furthermore, systematic energy management has been introduced in most public institutions and energy consultants encourage trade and industry to save energy by means of energy auditing schemes.

In 1996, new green tax legislation was introduced for trade and industry the revenue of which is to be recycled to business enterprises. Energy intensive enterprises can enter into agreements concerning energy efficiency measures which will allow them to receive a reduction in CO_2 tax. Subsidies may also be granted for specific projects leading to energy improvements in business enterprises.

To attain the overall objective of reducing CO_2 emissions, continual development of new, more energy-efficient technology is required. The Energy Agency administers the Energy Research Programme (ERP) that provides funds for research and development in a number of important fields including further recovery of oil and natural gas, combustion and gasification of biomass, development of large-scale, low-noise win turbines, and fuel cell and superconductor research.

There are currently four (4) power plants in Copenhagen. Surface heat is utilized for the power plants for process plants and to supply heat to the residential sector.

The current value added tax (VAT) on energy is 25 percent, with total taxes on energy amounting to a 50 percent level. Electric energy cost to the Danish household is currently at $0.20/kWh USD with taxes added. The average normal household consumes approximately 4,000 to 5,000 kWh per year.

Wind power currently accounts for 3 percent of the Danish power generation, with a target of 12 to 14 percent for renewable (solar, biomass) energy. Present windmill designs can generate up to 1.5 megawatts (MW) of power. The best overall combination of cost effective energy delivered to the Danish systems is produced from windmills in Denmark, and the purchase of hydro power from Norway. New plants built in Denmark consist of gas and straw fueled, with little emphasis on oil.

Currently, there exists seven (7) major generating companies, supported by 100 distribution companies. These companies feed into two (2) power pools. Privatization of the local distribution companies (LDC) is not on the foreseeable horizon. Denmark wants to focus on an electric system for the Danish, with little or no outside influence. Eventual liberalization of the Danish electric utility system will come about after Norway, Sweden, England, and the surrounding countries accomplish it.

STOCKHOLM

Meetings in Stockholm were held with Stockholm Energi. Representatives from Stockholm Energi, along with U.S. Embassy representatives from the commercial division, met with AEE delegates to discuss energy related matters focusing on deregulation, district heating, and district cooling as related to the city of Stockholm and Sweden.

Stockholm Energi primarily serves customers in the Greater Stockholm area and Central Sweden, and is one of Sweden's leading energy companies. The Company is broadly-based, including electricity, heating, cooling, gas, and electricity supply network operations. The unique production mix with electricity and district heating operations of the same size allows flexibility in adapting production in a cost-effective way to different weather conditions. Production of electricity in the district heating operation has become a competitive alternative in dry years such as 1996, when it was important to compensate for the low production of hydroelectric power. Stockholm Energi has its own facilities for production and distribution of electricity, heating, gas and cooling.

The ability to give customers the opportunity of complementing their energy solutions with the pro-environmental district cooling has provided Stockholm Energi with competitive advantages in relation to other power companies as regards to meeting customers' total energy needs.

Deregulation has meant that the Electricity Business Division is no longer restricted by the Local Government Act but may carry out operations without taking into consideration local government restrictions. As a result of the collaboration with Imatran Voima Oy, it will be possible to make use of competence in the Network, Heating and Cooling Divisions on a larger market, both nationally and internationally.

The Group has around 450,000 customers, the major part of whom are customers with domestic electricity. There are over 200 large consumers of electricity, over 3 GWh. Competition for customers became tougher when the electricity market was deregulated and some downturn in electricity sales in the Stockholm region was noted.

Network Supply operations have not been deregulated. The Group still has the sole right to transport electricity to all customers in the Stockholm and Avesta areas on its own networks.

Sales of heating have undergone great expansion during the year, and district heating has now around 65 percent of the heating market in Stockholm with around 3,500 customers. The greatest competitor for district heating is the local oil-fired heating boilers. The environmental benefits of district heating, operational reliability, and simplicity for the customer provide important competitive advantages.

The Group's oldest product, city gas, which has been supplied since 1853, is sold to 109,000 customers of which 102,000 are cooking gas market customers. Stockholm Energi's newest product, cooling, where the first delivery took place in May 1995, has around 80 customers. District cooling is primarily intended for commercial properties in central Stockholm. As in the case of district heating, district cooling's environmental profile, simplicity and operational reliability provide clear competitive advantages.

In February 1996, the Electricity Business Division introduced the product "Choice Power," allowing the customer to choose the method of production and account for their purchases of electricity between hydroelectric power, nuclear power or Stockholm electricity. During 1996, this has been broken down to 3,563 GWh of hydro power production, 4,964 GWh of nuclear power production, and 1,112 GWh of Stockholm electric power production. Of this complement, the following produced power was sold: 47 percent for hydro, 23 percent for nuclear, and 0.4 percent for Stockholm electric.

Hydroelectric power production was considerably lower than normal and 1996 went down inx history as an exceptionally dry year. Compared with Stockholm Energi's normal annual production of hydroelectric power of 4.8 TWh, only 3.6 TWh was produced, a reduction by 25 percent. The reduction in hydroelectric power was compensated for by our own more expensive thermal power.

During the year, a total of 10.4 TWh (10.5 TWh) of electricity was produced in our own and jointly-owned plants.

In the autumn, Stockholm Energi introduced the concept Mega Customer. Major electricity customers who purchase more than one of the Group's energy products can become so-called "Mega Customers." They then benefit from favorable contracts with their own customer representative, energy audit, monthly statistic, energy balance, and simplified invoicing procedures, among other things. The intention is to reduce the customer's total energy cost and to simplify administration in contacts with customers.

Stockholm Fastighetsagareforening was the first to enter into a Mega Customer contract for 1,401 properties.

In a normal year, approximately 45 percent of the electricity produced comes from hydroelectric power, 45 percent from nuclear, and 10 percent from local. Other power is produced in big-fuel, coal and oil-heated facilities. Approximately 75 percent of all energy produced today comes from sea water.

1996 was an exceptionally dry year with very little precipitation. Precipitation has not been so low since 1969-70. Due to this, the Group lost around 25 percent of the normal hydroelectric power production. Hydroelectric power was replaced by more expensive production at an additional cost of around SEK 250 million (approximately $34.1 million USD).

Prices on the Norwegian-Swedish energy exchange rose successively at the beginning of the year due to the initially cold period with low precipitation. When the expected spring flood was late and small, energy prices rose further. Price peaked early in the autumn, to fall subsequently as inflow increased and import of power took place, especially from Denmark.

In the 1990's, Stockholm Energi has aimed to create a balance in its own production of electricity between hydroelectric power and nuclear power. This strategy was followed up in 1995 when further hydroelectric power resources were acquired from Akzo Nobel. The consequences of a phasing-out of nuclear power will affect the whole energy sector.

Present electric rates for Stockholm converted to US $ equate to approximately $0.09/kWh (USD) for residential, and approximately $0.04 to $0.05/kWh (USD) for industrial and commercial sectors. Pricing of electric is broken down to approximately 1/3 for fuel cost, 1/3 for transmission costs, and 1/3 for taxes (varies for industrial and commercial sectors).

On January 1, 1996, the 90-year-old electricity legislation was replaced by a new electricity act, and the deregulation of the electricity market became a reality. Currently, the wholesale level is fully deregulated. Households are slowly being brought into the picture.

Sales amounted to 13.2 TWh (14.2 TWh) of which 1.8 TWh (1.8 TWh) were internal sales. The reduction largely depends on lower sales to other power suppliers.

The deregulation of the electricity sector led to tougher competition to the benefit of customers. It is to be expected that the exchange price and customer price will be more closely related, especially in the sectors exposed to competition. The margin between the price to the customer and the exchange price will diminish, as well as price differences between different suppliers.

The Nordic exchanges will be integrated which will result in the Nordic countries becoming a common market. The free electricity market, especially in northern Europe will develop successively in response to customer pressure. Through partnership and expanded competence, Stockholm Energi intends to participate in this European development.

Concentrated production of district heating makes it possible to use renewable sources of energy and thus reduce emissions compared with individual heating. During the past three years, Stockholm Energi has expanded its district heating market share by a further 10 percent. Local oil-heated boilers have been replaced by district heating or local heating solutions. During the same period supplies of district heating increased by the equivalent of the annual heating requirement for 25,000 detached houses.

For the third year running, district heating prices were unchanged despite increased taxes and rising oil prices.

New sales during the year amounted to the equivalent of SEK 100 million/year (250 GWh/year) in income. This expansion took place primarily in Sodermalm, and through new establishment primarily in the southwestern suburbs.

In all, sales amounted to 6,981 GWh (5,993 GWh) including deliveries to Sollentuna, Jarfalla and Avesta with 440 GWh (387 GWh). Heat pumps, pellets, bio-oil, and refuse fuel accounted for 60 percent of the energy supply. Coal, oil and electricity accounted for the remainder. Electricity production amounted to a full 1,627 GWh during the year, of which 515 GWh came from production in condensing power plant. The high proportion of condense production was caused by high electricity prices due to the dry year.

Simplicity is created primarily in relation to customers. Invoices via EDI are already a reality and more and more of our customers will be able to read consumption statistics etc. via Internet. All customers were given the opportunity of becoming Mega Customers of Stockholm Energi.

In regard to gas supply, the Business Division of Stockholm Energi offers a competitively-priced range of gas products to customers in the Greater Stockholm area. City gas sales amounted to 396 GWh, an increase by 34 GWh. Marketing activities are focused on continuing efforts to make new sales to real estate owners. New agreements equivalent to deliveries of 26 GWh were established.

During the year, SE Gas was given the assignment of creating good conditions for continued gas operations. Active efforts to reduce the annual loss of cooking gas customers, and to develop the home-heating market were initiated. Real estate owners can now, for instance, enter into agreements on residual value guarantees when replacing old gas cookers by new.

The number of customers was 109,000 (113,000) at the end of the year, of which 101,500 (105,000) were on a cooking gas tariff. Customers are in Stockholm, Solna, Sundbyberg and Nacka. Stepping up the tendency for customers to move away from gas cookers is of decisive importance for SE Gas's future profitability. The Business Division has therefore started active effects to maintain its place on the cooking gas and home-heating markets. The aim is to reduce the loss of gas cookers from 4,000 per year in 1996 to 2,000 by 2000.

City gas production takes place in a cracked gas plant. 41,500 tons of naphtha were used as raw material in 1996.

New sales of building heating are continuing. Use of gas as a vehicle fuel is being developed further. Within the framework of the EU-supported ZEUS project, 180 vehicles will be powered by biogas.

District cooling, with the aid of cold sea water, has been successfully introduced as a new product and a strong complement to Stockholm Energi's traditional supply of energy. After over two years, district cooling has achieved a market share of 35 percent on the cooling market in the distribution area.

The year's new sales 26 MW were made to over 40 customers. The connection of new customers meant that local refrigeration plants containing around 7,000 kg Freon could be eliminated. At the turn of the year, the total connected capacity was 52 MW (26 MW) and the energy supplied during the year reached 34 GWh (17 GWh). The city system includes a connected capacity of over 120 MW.

Stockholm Energi's unique district cooling investment has met a response throughout the world. The concept for establishment of district cooling can be introduced both in Sweden and abroad.

OSLO

AEE delegates met with ENRON Nordic Energy in Oslo, Norway. ENRON is directly involved with the energy commodities market and NORD POOL. ENRON buys and sells energy contracts for various industrial and commercial customers as determined by the unit cost of energy set by time of day and daily pricing of the Pool.

Norway's total production of electric energy in 1995 was reported at 123.2 TWh, which is an increase of 8.8 percent from 1994, and 1.4 TWh over the old production record from 1990. Exports of electric power (in 1994) amounted to 4,836 GWh, while imports accounted for 4,968 GWh. Total hydroelectric production capacity with normal supply of water is estimated at 112.3 TWh.

Typically in a dry year, the coal-fired plants are the last to be brought on, and in a wet year, the nuclear plants are the last to be brought on. This is typical for the Norwegian market of Norway, Sweden, Finland, and Denmark, and the power grid system. Total energy consumption for the Norwegian countries is approximately 370 TWh.

There are currently no nuclear power, coal or gas fired energy plants in Norway. Ninety-five percent of the electric supply (in excess of 118 TWh annually) is based on hydroelectric power produced in more than 600 power plants throughout the country. Norway is considered to be the world's largest consumer, per capita, of electricity. Most of this energy has been generated, controlled and distributed by Statkraft and Statnett Grid, two GON-owned entities with an annual investment level of about USD $200 million. However, hydroelectric energy has gradually been liberalized from a strict local monopoly status, and the various power plants are now individually permitted to compete with each other domestically and on international markets.

From the total 1995 production of 27,555 MW, Statkraft supplied close to 30 percent; municipal and local government utilities supplied 55 percent, while private and industrial power plants accounted for 15 percent.

The annual per capita electricity consumption in Norway at present is approaching 25,000 kWh. The average price for electricity to regular consumers was in 1995, NOK .55 per kWh, ($0.086/kWh, USD), while average price for heating oil was reported at NOK 2,55 per liter (NOK 6.40 equals USD 1.00). Norway's energy rates to the end-user are typically broken down to: 1/3 wholesale cost + 1/3 transmission + 1/3 taxes. All transport and distribution lines are operated by Statnett (Norway's state owned Power Grid Company) controlling a grid of more than 200,000 km.

GON, Norway's state owned power grid company, has opened up the local markets to competition so that the nearest power plant does not have its own area automatically guaranteed with monopoly pricing. Local power plants are now delivering electricity to various organizations and customers located in Oslo or elsewhere in the country. Norwegian authorities claim that Norway was the first country in Europe providing free competition among the various domestic energy producers. To date, even with deregulation, no real net savings on energy have been realized at the end-user level. Profits have been basically observed at the transmission level.

Power-intensive industry such as aluminum works, ferrosilicon works, and wood processing plants currently account for 25 percent of the power consumption in Norway. The rest of the industrial sector, the service industry and transport consume 45 percent. The remaining 30 percent is consumption by households.

Noting that hydroelectric power accounts for virtually all electricity generated in Norway, there is a general consent that a highly developed country cannot base its power supplies exclusively on the weather. The Government has decided to build two gas fired power plants at an estimated value of approximately USD 500 million on the west coast of Norway. Electric energy has always been regarded as an important export potential, particularly to Norway's neighboring countries. Abundant Norwegian natural gas resources in the North Sea is now scheduled to be pumped into two 350 MW gas fired power plants. The project recently obtained approval from the Norwegian parliament and is managed by Statoil (Norway's state-owned oil and gas producer), Statskraft (Norway's state-owned hydroelectric power utility), and Norsk Hydro (gas producer and industrial flagship, -51 percent controlled by the Norwegian state). Each company has one third ownership in the gas fired plant. The two plants are scheduled to be in operation by 1999 and 2000 respectively.

CONCLUSIONS

Since the implementation of electric deregulation/privatization in England in 1990, the impact of full deregulation in various countries visited and reviewed, has yet to be realized. While much has already been achieved since privatization, many challenges remain in England and the Scandinavian countries.

Some of these challenges, future areas of change, and general comments include:

- On the generation side, efforts must be made to insure that competition continues to drive down costs and that these savings feed through into further price reductions for the customers.

- In general, the transmission portion of the electric utility industry must be deregulated to fully realize the impact of end-user cost reductions.

- In England, the regulated parts of the business, electricity companies will have to meet more demanding targets set by the Regulator, which are imposing real price reductions from year to year.

- Privatization does not appear to be in the near picture for the Netherlands.

- Eventual liberalization of the Danish electric utility system will come about after Norway, Sweden, England, and the surrounding countries accomplish deregulation/privatization. Denmark's focus is on developing and enhancing the electric system and infrastructure for the Danish, with little or no outside influence.

- In Sweden, currently the wholesale level is fully deregulated. Households are slowly being brought into the picture.

- The Nordic exchanges are shifting towards an integrated exchange which will result in the Nordic countries becoming a common market.

In summary, the expansion of electric deregulation will create expanding opportunities which will result in enhanced energy cost savings to the end-user. Controlled deregulation of the electric utility industry would be a prudent course to follow. This would still allow for end user savings to occur, and focus on not diminishing the overall integrity of the electric utility system. As we are about to enter the twenty-first century, we are beginning to see the future of the energy industry and its impact on our world environment.

CHAPTER 4

PROPOSED NEW ENERGY LEGISLATION: HOW WILL IT IMPACT POWER MARKETING TRANSACTIONS?

As a transactional attorney poised to follow the new power markets and assist in their financing, I follow the legislative debate not from the perspective of the absolute public policy merit of positions (which generally are rhetorical anyway), but from what they will mean for entrepreneurs seeking to exploit the potential of open access via sales or via aggregation.

While that means, of course, that I (like most of you) am fundamentally in favor of open access, it also means that I keep my eye cocked on the implication of certain key questions. What I would like to do with you is consider:

(1)   What the overall approach of the three archetypal bills before Congress are to the generic questions of interest to those involved in power marketing;

(2)   How interested parties might meld their self interest with respect to these questions with the titanic struggle of other interests reflected in the legislative fray; and

Printed in *Cogeneration and Competitive Power Journal*, Vol. 12, No. 4, by Roger Feldman. Copyright 1997 by Bingham, Dana & Gould.

(3) Identify what special key questions are presented by these
     bills and what additional statutory guidance they might seek
     in this regard.

As a baseline, let's begin by defining the power marketer's
legislative self interest. Recognizing that power marketers are not
a uniform bunch there are a few keys to their success.

•   First obviously, market homogeneity making feasible the
    treatment of power as a commodity freely transportable and
    tradable without necessary barriers in interstate commerce.
    Read, in legislative terms: State versus Federal jurisdiction
    and treatment of stranded costs;

•   Second, protection from de facto inability to reach markets
    because of practical economic or operational constraints on
    doing so, or newly imposed regulatory requirements. Read,
    in legislative terms: market dominance through anti-competi-
    tive actions or through affiliate ownership arrangements;

•   Third, protection of the feasibility of those activities which
    enhance the long-term competitive position of those in the
    power marketing business. Read: focus on key business
    trends which power marketers are now recognizing as the
    competitive parameters of the market become clearer. These
    include:
    —   convergence of gas and electricity sales
    —   combination or competition with energy management
        service provision
    —   Flexibility in aggregation arrangements
    —   Niche emphasis on utilization or aggregation of renew-
        ables

In considering the ramifications of the bills against this
baseline, it is useful to keep in mind their origins and different

focuses. The **Schaefer Bill**, introduced last year, represented a bold thrust toward deregulation, but one designed to offer placating incentives to electric utilities and to holding companies. The just introduced **DeLay Bill** is consistent in spirit with Schaefer, but also is designed as a warning shot across the bow of the recalcitrant utilities. The **Bumpers Bill**, from the other side of the aisle, supports deregulation, but reflects additional environmental and populist concerns. Some synthesis of the three bills would be ideal, with a few additional twists.

Now, let's look at the key generic policy denominators of these bills which affect power marketing.

(1)   First, the extent of substitution of federal directives for state action. The underlying conflict here is between open access proponents, anxious both to press their case and avoid the need to deal with a crazy quilt of jurisdictional responses to open access and a variety of supporters of the contrary position, which include not only higher cost "just go slower" utilities; but also legislators from lower cost power states; certain consumer groups; public power systems; and, of course, state regulators.

The **Schaefer Bill** seeks to straddle this issue, by providing the states with an opportunity to conduct proceedings to establish nondiscriminatory retail access by a date certain, as reflected in proceedings which include mandatory consideration of specified principles. Nonregulated utilities are afforded this opportunity as well. In the event of state failure to implement retail access in accordance with these principles, FERC is directed to step in and by December 15, 2000 effectuate customer choice.

The proceedings designed to provide the state portals to full deregulation and the cessation of state price regulation are required to insure universal power availability, reliability, energy efficiency, conservation and environmental programs—*and* cost recovery of stranded investment and PURPA contract costs. The potential complexity could exceed the old PURPA proceedings.

There is also a contemplation of a FERC findings separating

its jurisdiction over unbundled interstate retail transmission and continuing state jurisdiction over local distribution.

DeLay's proposed legislation would delete the craftsmanship of the Schaefer Bill—leaving regulatory authorities only with deference in certain decisions relating to electric service, and obliterating any distinction between regulated and non-regulated utilities.

It also articulates final "consumer choice" conclusions which otherwise might be more finely tuned in state proceedings: a ban on exit fees, subsidies or other penalties on exercising right of choice; a ban on restriction of alternate choices.

State authority is left over distribution systems; protection of safety and reliability; and of consumer rights. States are also left with a variety of responsibilities of interest to power marketing specialty firms: interim rates; continuation of universal service; conservation programs and initiatives; consumer choice with regard to renewable energy; research and development. In other respects, FERC is directed to provide for nondiscriminatory prices with respect to distribution (as well as transmission).

The Bumpers Bill would mandate retail access in 2003, unless earlier instituted by state governments. More expansively than the other alternatives, it continues consistent state local distribution and retail transmission regulation. Bumpers would permit stranded cost recovery where retail energy regulation requirements are met, subject to broad general guidelines and specific multistate utility company stranded cost rules, but insulated from retroactive prudence review. It also would specifically permit nuclear decommissioning cost recovery.

More than its Republican counterparts—which focus almost exclusively on market deregulation—(except in the area of Renewable Energy Credits), Bumpers also makes specific provisions for "reregulation" of the deregulated transmission sector. It makes recommendations, for example, for transmission regions; standards for and regulation of Independent System Operators through Regional Transmission Oversight Boards. We may hear more about these requirements.

From the power marketing standpoint, specifically, the resolution of the state authority issue ought to be to make retail access the law of the land, right away. But the straightest available line to this point politically may well be: (1) specific federal acknowledgment of state authority to implement retail wheeling; (2) Emphasis on the development of state consistency, through establishment of more detailed federal standards than currently are under state consideration. The possibilities of acceptance are enhanced.

With respect to the treatment of the stranded cost issue, many power marketers may applaud the DeLay approach: if implemented, it certainly would take down several of their competitors; permit greater differentials in competitive rates offered by power marketers; and create resultant incentives for major power flow transactions.

While all of this may be true, it seems very unlikely to happen, given the institutional strength of the utilities; utility willingness to accede to restructuring, at least superficially; and the welcoming of open access. On the other hand, right to stranded cost securitization—a matter of great interest to utilities—might more productively be made an issue of permitted degree; a function of timing in moving to open access; and subject to conditions of transmission system management raised by Bumpers

(2) A second generic basis for distinguishing among current legislative proposals is their treatment related to assurance of removal of market imperfections. Principally, this will relate to two issues: treatment of market power, and the form and scope of dismantling of the Public Utility Holding Company Act (and PURPA).

The Schaefer Bill, whether from naive or disingenuous faith in how private firms actually behave in free markets, or fundamental aversion to the recrudescence of intrusive regulation, basically does not focus squarely on the market power issue. It does make clear that the antitrust laws are not superseded, but for the most part its focus is on assurance of the removal of governmental barriers to competitive market entry.

Perhaps because of its intention to pressure utilities to reach closure on open access, or perhaps because the full ramifications of the utility merger trend were coming to light by the time it was developed, DeLay would direct FERC specifically to ensure that present and future utility exercises of market power do not impair the objectives of open access. The proposed authority is very broad, extending to restrictions, under certain circumstances, on sales at market prices and divestiture orders. Bumpers would go even further, focusing an entire statutory Title on "Competitive Generation Markets," providing FERC with authority to deal with situations "inconsistent with effective competition among retail and wholesale electrical providers.

It is likely that efforts to, in effect, make FERC a free market policeman will encounter serious political resistance. The issues already faced in the merger context would pale by comparison in number and scope. On the other hand, market strength abuse ought to be an area of considerable concern to power marketers, given the rapid transformation of the form of the market to one in which large utilities will be servicing and seeking to retain large numbers of customers.

It is somewhat ironic that at the same time as markets are slated to be made more competitive, there is a perceived need by some to give FERC the general type of oversight powers which the SEC has exercised with respect to holding companies under PUHCA—which all of the proposed bills contemplate repealing. The logic in the latter case, is that the need for PUHCA is obviated if the competitive circumstances necessitating its imposition are removed. The politics are that the same utilities which are being asked to cede open access—and basically are opposed to it—are also, however, anxious to see Holding Company Act repeal.

Schaefer devotes a separate Title to PUHCA repeal, picking up many of the provisions found in the special purpose legislation which was introduced last year. Basically, it contemplates that the

PUHCA shackles will be removed when retail access is established in all states in which the holding company operates. Authority is provided for federal access to books and records, and for state access as well. Continued federal and state jurisdiction is provided over affiliate transactions. The DeLay Bill continues parallel deregulation provisions, but without the very important books and records provisions. As might be expected, Bumpers not only contains analogues to the Schaefer Bill provisions, but also would extensively empower FERC to police interaffiliate transactions.

PURPA repeal is an element of all three bills, to be triggered upon full realization of open access. The proposals are focused on preservation of old contracts. Only Bumpers explicitly purports to protect against mandatory downward negotiation of existing rates.

A possible additional condition to these PURPA repeal provisions, which many power marketers would favor, would be the preclusion of utility participation in distant territories which have been deregulated, unless their own territory has been deregulated at the retail level as well. This would seem particularly relevant given the linkage of some power marketing efforts to projects still enjoying or in need of PURPA-type benefits.

TRANSACTIONAL NICHES

(3)   While the battle swirls around the larger generic legal issues just discussed, power marketers would do well to assure that the new legislation also serves the purposes of enabling them to continue to develop and expand their own transactional niches: convergent energy supply services; interface of energy management/ service and power marketing.

(a) First, power marketing and natural gas marketing (not to mention the marketing of additional fuels) are becoming a single (sometimes national, sometimes regional) enterprise.

Schaefer recognizes this in its PUHCA authorization—which requires as a condition of the lapse of PUHCA jurisdiction over a holding company, that its gas utility customers also have the benefit of retail open access in the gas field as well. In all other respects, its focus is on electric retail access. DeLay does not deal with the issue. Bumpers' more extensive concern with the potential unfair impact of market competition extends to "natural gas utility company" acquisition, to which it would apply a broad "public interest" standard to permitted acquisition. It also has an explicit anti-cross subsidization provision.

A key strategic issue for many involved in energy marketing is whether it will be sufficient for them to continue their reliance on applicable gas and electric deregulation provisions fitting together, or whether the interface of convergence and deregulation should be acknowledged specifically, so that they can compete effectively in offering energy services. It could be the predicate of a proactive or a defensive strategy, depending on the power marketer.

(b) The issue of interface of power marketing with energy management service activities presents a similar type of two edged sword. Some power marketers have offered ESCO services and vice versa. Some of each have emphasized the superiority of their respective approaches.

The Schaefer Bill captures in its definition of "retail electric energy service" the full panoply of activities which may occur as a result of retail access. It includes billing and metering, "electric management services" for ultimate consumers and other electric service alternatives to those offered by applicable utilities. State open access implementation proceedings are required to address energy efficiency. The DeLay Bill is as expansive in its definition, including virtually all services other than transmission and distribution. Bumpers would seem to be more narrow in the markets it opens: it applies to "ancillary services sold for ultimate consumption." Remember, the narrower the definition the smaller the explicit window for deregulated utilities.

Markets for power marketing are created through, or in conjunction with, effective energy efficiency activities. The issue of whether E.M.S. services are properly treated as an unbundled component of supply or merely an ancillary activity would benefit from closer delineation. For example, it impacts the issue of the extent to which utilities can use energy management service provision as a basis for protecting their market share. It could also impact the issue of a potential split between state regulation of ESCOs and federal regulation of power marketing.

The issue is indirectly, of course, related as well to the interface between power markets and customer groups—the type of "aggregation" which has begun to emerge nationwide. Aggregators may facilitate their operations by engaging in energy management as well. Interestingly, while the Schaefer Bill is a consumer driven statute, focused on removing barriers to consumer sales, it does not specifically make provision for the rights to consumers to form cooperative purchasing arrangements.

Bumpers, by contrast, specifically acknowledges the existence and permissibility of aggregation, but only if the members are in a state "where there is retail electric competition." While presumably this was not meant to be limiting language, it could have that limiting effect if—for unrelated reasons—other disputes are proceeding as to whether there is retail electric competition in the jurisdiction.

The treatment of renewables as a specially favored source of electric energy could have a positive benefit to the limited population of power marketers/aggregators who wish to focus on renewable energy supply. It could, however, result in an additional burden for those marketers also directly or indirectly in effect are compelled to traffic in renewable energy.

Under the Schaefer Bill, each "electric generator" must meet a quota for renewable energy—not including hydro. The applicable percentage would progress by increments from 2% in 2001-2004 to 4% after 2010. Renewable energy credits will, however, be tradable, and therefore available to be purchased to meet these requirements. Renewable PURPA contracts will be available to be credited to purchasing utilities. The DeLay Bill is silent on the use of renewables—probably reflecting its aversion to the imposition of new regulatory disciplines. Bumpers proposes a comparable but more elaborate renewable energy credit package. It also provides for the accommodation of additional requirements of state renewable energy programs as well.

CONCLUSIONS

What, then, are the prospects for power marketing under the new legislation? Until the full outlines of the legislative compromises are visible, it will not be entirely clear how rapidly retail access and the opportunity for power marketers will emerge. As with wholesale wheeling, it may be that actual market transactions serve to define the considerable fuzziness which characterizes the standards. It would be a mistake, however, to assume that this will automatically be the case. Special provisions affecting the energy power marketing business have not been crafted carefully. They deserve focus by industry players.

Power marketing based transactions, and transactions launched to take advantage of power marketing, need to be planned now with the possibility of the legislative scenarios in mind. Open access is a foundation of market opportunities, but not, by itself, for product differentiation.

CHAPTER 5

NEW TACTICS AND TECHNOLOGIES TO MEET THE COMPETITIVE UTILITY ENVIRONMENT

INTRODUCTION

A new age is dawning for lower-cost energy use and supply. The deregulation of the electric industry is creating new pricing options that will change how we evaluate cost-cutting energy alternatives. As competition begins, smart users will grasp these opportunities and press for greater innovation on the part of marketers.

Energy users can best navigate these choices by:

•   understanding the concepts inherent in deregulation (such as transmission constraints)

•   influencing the deregulation process (which does not end when markets first open)

•   learning to use new analytical tools (such as load profile analysis)

Presented in *Strategic Planning for Energy and the Environment*, Vol. 17, No. 4, by Lindsay Audin, CEM, CLEP, IES

- applying new technologies (e.g., wireless automatic metering)

- being as creative as possible (because marketers won't be).

AT&T VS. MCI: A PARADIGM

To understand how electric utilities are transforming, think about how long-distance phone service has changed. When AT&T was forced by a federal court to divest its divisions, long-distance and local services were separated, and new providers such as MCI and Sprint became household names. After 13 years, that industry is still not fully deregulated (local service is still generally a monopoly), but during that time long-distance use has nearly quadrupled, while the average price of a long-distance minute fell by more than 50%.

To satisfy consumer demand for communication services, a vast array of new technologies was also born. How many of us anticipated the home fax machines, cellular phones, pagers, and on-line services that would result from a single court order? While we can also expect the cost of power to eventually fall, the future of electricity similarly holds much more than price reductions.

FACTORS IMPACTING POWER PRICES

Electricity is generated by utilities and independent power producers (IPPs), both regulated to some degree by state public utility commissions (PUCs), and then transmitted through high-tension lines criss-crossing North America in a giant network. These lines are owned by utilities and regulated by the Federal Energy Regulatory Commission (FERC). Once voltage is stepped down at substations, power is distributed through local utility-owned lines and meters regulated by PUC's.

With the exception of rural co-ops and municipal utilities (which are controlled by local governments), PUC's determine how to distribute these costs to end user classes. Most of our bills break out only charges for electric consumption and demand (and perhaps a fuel charge), but the true cost of power includes many other components, including transmission, distribution, and a variety of ancillary services (such as voltage support, spinning reserve, and load following). Bills may also include taxes, social programs, and other charges that are not apparent to end users.

To develop the prices we pay, the PUC's apply a standard based on the utility's costs for providing a service, plus a guaranteed rate-of-return to ensure a ready supply of investment capital. All of these costs and profits are "bundled" together to create tariff pricing. While theory dictates that charges should be based on the true cost-of-service, politics and other pressures often result in cross subsidies in which one rate class (e.g., industrial) is charged more to contain prices charged to another (e.g., residential).

While the electric rates we pay are controlled by PUC tariffs, utilities and IPPs buy and sell electricity among themselves, and such wholesale prices vary with time, climate, power plant outages, fuel prices, and other factors. The base cost of power seen by a utility is therefore a mix of its own generating costs and the price it pays for electricity delivered from other power providers through the transmission system. This base cost is subject to commodity market conditions usually not visible to end users, and increasingly influenced by factors such as commodity trading techniques (e.g., futures and financing plans), user load profiles (such as real-time pricing), and transmission system constraints (that can drive prices up to the highest local generating cost).

THREE GENERAL RELATIONSHIPS

There are three general relationships that clarify how new techniques both interact and can be applied to control energy pric-

ing. They are:
- time of use - load variation - hourly pricing
- generation - transmission - natural gas options
- load shaping - financing methods - user technologies.

Time, Load, and Price

As they become more time-sensitive, deregulated retail power prices will begin to vary like those at the wholesale level; i.e., hour-by-hour. Since one's demand for power generally changes during the day and the week (and by season) we can expect the average price to also change with time and use (unless controlled by other factors, discussed below). As a result, load profile shapes (i.e., a graph of power versus time) will influence pricing, with flatter profiles generally having a lower average cost. Utilities generally typify such patterns via load factor, defined as average demand divided by peak demand. A high load factor would indicate a flattened profile while low load factors would occur where a peak demand is relatively brief, and is surrounded by much lower demand during the rest of the day. While the demand for power in most non-industrial buildings varies with time, it usually does so in predictable patterns. Knowing the shape of your typical daily load profile can often reveal ways to cut the present and future cost—and price—of power, while also helping your power supplier offer the best and most secure pricing.

Like Sprint's "dime-a-minute" long distance rate, marketers will likely offer highly simplified rates that smooth out such time-based price variations, but subscribing to such options is unlikely to yield the lowest average power costs. Rates that vary widely over time may provide the lowest average price, and techniques that cut, level, or shift peak demand will help reduce those prices.

Transmission, Generation, and Natural Gas Options

In some areas, peak loads exceed transmission capacity many hours each year. When low-cost power can't be brought in, prices could be bid up to the highest local generating costs. A good example of such constraints appeared during the early hot spell of June 1997 in the PJM (Pennsylvania-New Jersey-Maryland) power pool. While daily bulk wholesale generation prices (which make up 30% to 60% of most bills) generally don't vary from one end of the pool to another by more than $.01/kWh, June saw *variations exceeding $.13/kWh* when transmission constraints blocked cheap power from reaching high-cost areas[1].

In some urban areas with older power systems (such as New York and San Diego), transmission constraints could yield similar results. Such areas with constraints are sometimes called "load pockets" during the period of constraint (which may exceed 1000 hours a year). In the United Kingdom, which uses a national power pool supplied by deregulated generators, power suppliers have also found ways to "game" the system to purposely congest transmission, thereby driving up the price of their product[2].

To address such possibilities, some energy marketers have begun promoting new local generation (or cogeneration) facilities, either at customer-owned sites or through re-powering of obsolete utility plants inside the load pockets. Natural gas generators with very low emission levels have become quite cost-competitive for both peak shaving and as base load power, opening the door to competition during transmission constraints. Similarly, a variety of technologies (discussed below) exist to reinforce existing transmission systems. A recent study[3] found that a small investment toward improving transmission capacity on one major power line in California could have a major impact on limiting summer power prices.

Substituting natural gas for electricity during peak pricing periods can also impact power prices. A variety of technologies exist for using gas to directly provide horsepower, cooling, air compression and other power-intensive needs. Such convertibility will create truly interruptible energy rates, allowing clever end users to contract for both interruptible gas and power, attaining the lowest possible energy prices. Under these circumstances, transmission, generation, and natural gas options will compete with each other, driving all prices down over time.

Load shaping, Financing, and User Technologies

A variety of choices are emerging to shape loads in advantageous ways. While most have been around for some time, deregulation will allow marketers to help end users gather—and isolate—their loads more readily through metering and contractual means. In the end, there is little (if any) impact on the service (e.g., chilled water) supplied. Instead, each of these options reshapes loads (as portrayed to utilities and regulators) without major alterations to end user facilities.

Load Shaping

Coincident metering often cuts the average cost for power at facilities when many meters on different accounts serve one customer. At Columbia University in New York, for example, gradual expansion without attention to energy costs resulted in one property having several dozen accounts and meters. Each account peaked at a different time, but (due to tariff construction) peak demand charges were treated as though all buildings had peaked at the same time. By combining the accounts under one master demand meter, the average cost of power was cut by over 10%. Such combination will become easier under deregulation as usage (for the power commodity, as versus transmission and distribution) can be contracted under one account.

Load isolation, while not favored by utilities, can allow an end user to segregate loads with poor load factors (such as electric chillers) that would be cheaper under utility tariff-based power than under market-based power, thereby reducing the average cost for all loads.

Demand cooperatives (promoted by Planergy, Inc.[4]) are a way for end users to work together (typically through an organizing vendor) to obtain lower utility rates by pooling interruptible loads and agreeing to curtail them when requested by the utility. Creating such a cooperative distributes the need for interruption so that only a few loads are interrupted at any one time, but all participants gain some benefit.

District-wide systems serve multiple customers with (for example) chilled water from a central facility that, in effect, transfers many individual electric chiller loads (with low load factor) to a high load factor central facility that uses both electric and gas-driven units to minimize total cooling costs. Existing chillers may remain, but are either bypassed or cycled through a connection to a common chilled water loop serving numerous facilities.

Bill consolidation allows many accounts held by one customer to be gathered for both coincident metering and attainment of cheaper energy block load rates previously beyond any one account.

Aggregation involves the gathering of different customer accounts through a third party for purposes of bulk power purchasing, coincident metering, bill consolidation, transmission capacity reservation, and expert load analysis. Aggregators work along the same lines as MCI, buying co-ops, credit card handlers, and other organizations that compete for the privilege of bringing many end users together. All provide lower prices through bulk purchasing and handling. Present-day utility customers are, in effect, already "aggregated" into rate classes, but only to the point of developing a rate based on an assumed typical load profile.

Financial Tools

In similar ways, various financial tools exist that can often cut or levelize costs more readily than engineering solutions. Both marketers and financial firms are providing access to financing that ensures prices do not vary beyond predetermined levels. To take advantage of these opportunities, end users need to understand the following basic concepts:

- Electric futures (available on the West coast, and soon in many Eastern areas) are traceable contracts for monthly blocks of firm power, purchased in advance of need, that will

later be provided during normal business hours at a predetermined price. They are, in effect, promises for supplies of
power, though there is not necessarily a guarantee of delivery
unless accompanied by a secure transmission arrangement.

- Options (typically in the form of "calls" and "puts") are contracts between suppliers and marketers (and thus end users)
that allow a user of power to know that he can "call" for
power and be ensured supply, or that a seller of power can be
ensured a buyer when he "puts" out an offer to sell, both at
predetermined prices.

- Payment plans and weather insurance have been offered to
low-income customers by utilities for years. Now, however,
marketers are ensuring levelized (or predetermined) monthly
bills (not just pricing) through risk management techniques
that involve financial and load analysis tools. While some
monthly bills may be higher than in prior years, other
months will be lower, and the annual total will be confined to
a narrow range. Such plans are often complicated, requiring
careful analysis.

- Tolling allows an end user (or his designee) to provide a generator with boiler fuel (typically natural gas) in trade for electricity at negotiated, non-tariff, rates. This process is common
among power marketers also trading natural gas, and is used
at times that utilities have excess generating capacity that can
provide power into another utility's territory.

For a more complete discussion of energy risk management tools,
readers are referred to http://www.powermarketers.com, the
Web site of the Power Marketers Association.

User Technologies
      While all of these options provide end users with choices for
controlling energy pricing, their impact can often be maximized

when used in conjunction with new technologies. Just as we have seen an explosion of choice in communications, the future will see a variety of ways to create, store, and manage power. Many are already being offered, or are in the prototype stages of development.

Metering. Marketers are using more sophisticated power metering as a sales tool to ensure that their clients' power costs are minimized. The new standard in metering involves wireless communications, and automatic hourly, quarter-hourly, or real-time monitoring. Human meter readers, obsolete for many years, are disappearing as these cost-effective systems are installed. Such systems allow better load control, more accurate power nominations, better pricing and a reduction in theft and tampering (a major problem at the residential and small commercial level).

Software

Computerized building simulations and load analyses have taken on new prominence as tools for predicting and flattening load profiles. Analyzing short intervals (1/4 hour) has become essential to maximize savings through tighter load control and on-site power supplies. Names and acronyms such as PEDA, RBOSS, and PowerManager are being heard as marketers offer new services to cut electric bills and power pricing.

Energy Management Systems (EMS)

As responding to real-time or market-based pricing signals becomes a common way to attain savings, an EMS takes on new importance for controlling variable loads, such as fans, pumps, DHW heaters and chillers. When tied into the metering and software tools mentioned above, the load managing power of an EMS can be greatly enhanced.

Power Storage Devices

While many power practitioners and regulators continue to assume that power cannot be cost-effectively stored, flywheel power storage units are now in use as uninterruptible power systems (UPS) to supply "clean" power, and small (2 kWh) units act as backup power for cable TV systems. Larger units (over 12 kWh) are being prototyped as peak shifters/shavers for buildings. Chemical battery technology has advanced considerably, and will also play a part. A storage system that gets "filled" with cheap off-peak power at night (when there are no transmission constraints) and "empties" to flatten peak loads during the day, could be an instant money-maker for a smart marketer and/or end user.

Distributed Generation

The ability to generate power in a pinch has always been useful. New small (C100 kilowatt) modular gas-fired turbine generators[5] can also cut billing for peak demand by operating in parallel with utility power, or (to avoid backup charges) by feeding dedicated loads. Using ceramics and few moving parts, the low emissions of these devices create an intriguing option, while natural gas fuel cells with extremely low emissions) have already racked up an impressive operating record. Once prices on these devices come down, watch marketers and end users grab them up to minimize stranded cost payments. Smart utilities may also invest in them to minimize transmission constraints and/or provide competitively priced localized power during such constraints.

Transmission and Distribution (T&D) Networks

Since market-based power is cheap only when transmission is available to move it, pressure for new or more robust transmission will increase as large price differentials between adjacent areas become visible. Thyristor-based switching of high-voltage loads on T&D systems can raise the effective capacity of existing transmission lines. Such options are among a family of Flexible AC Transmission System (FACTS) improvements under development or deployment. Even new types of underground high-voltage cables[6], designed for use in transmission-constrained urban areas, are being rolled out to meet the expected demand for beefed-up transmission.

Gas-Powered Motor Drives. Natural gas-driven devices, such as gas engine-powered air compressors and chillers, have replaced electric motor-driven units in industrial and commercial facilities, cutting their peak demand.

Advanced HVAC Systems. Chemical desiccants dehumidify outside air using natural gas, thereby cutting peak electric chiller loads. This process is already common in new buildings with large outside air loads (e.g., hospitals) and industrial processes (such as air compression). For smaller facilities configured around rooftop units, Entergy has been offering a super-high-efficiency replacement unit which takes advantage of several refrigeration engineering innovations[7].

Many other options are either in the queue or already being sold. Try to imagine what the "fax machine" or "cell phone" of tomorrow's power industry will look like. As we have seen in other industries, the combination of several new technologies often results in devices few of us could have imagined only a few years ago.

Who Offers these Options?

Accompanying the profusion of technical choices is an ever more bewildering expansion of vendor choices. Even as merger/acquisition mania creates new firms out of old ones, most energy services providers (ESP's) continue to fall into a few distinct groups.

- Unregulated utility subsidiaries; e.g., Cinergy, Southern Electric Corp.

- Independent power providers: e.g., Sithe Energies, Calpine Inc.

- Mega-wholesalers: e.g., Enron, Duke Energy

- Equipment vendors: e.g., Johnson Controls, Honeywell

- Gas marketers: e.g, Eastern Energy Marketing, Colonial Energy

- Existing ESCOs: e.g., Xenergy, EUA Cogenex

- Aggregators: e.g., New Energy Ventures, Wheeled Electric Power Co.

- Financial firms: e.g., Goldman-Sachs, Merrill Lynch

And, of course, local utility distribution companies (UDCs) are also trying to retain their load, despite claims that they are becoming neutral deliverers of others' power and gas.

The College of Power Knowledge

How does one cope with this continuously changing panorama? Fortunately, both the advancing energy industry and other innovations are providing some of the means to do so.

Getting Up to Speed. There's no need to enroll in college (none of them teach this stuff anyway). Start by learning the "lingo" and concepts. Most PUC's provide readable summaries of their decisions (both on paper and on their Web pages), and a variety of newsletters and free magazines are available to keep abreast of the latest changes (see appendix 1 for a list). Computer-savvy managers can "surf" informative Internet sites for even quicker access (see appendix 2 for another list).

Attending a conference focused on competitive energy issues can be very helpful to get your questions answered (see appendix 3 for a list of seminar providers). Such events are also a good way to make useful contacts. Be sure the event you would like to attend is not geared mainly for marketers, however, or you may end up both disappointed and confused. Local trade associations often sponsor panel discussions on deregulation issues, or are open to holding them if interest is expressed by their members.

Speeding Up the Process. You may already belong to a trade or professional organization that is (or could be) taking action toward deregulation. Several local BOMA (Building Owner's and Manager's Association) chapters, for example, are already actively pursuing power issues. To properly represent your company's interests, membership in a customer group—or working through an energy "partner"—(i.e., a consultant or marketer) involved in rate proceedings can also be of great value. Your PUC can provide lists of groups that have intervened in deregulation proceedings. On the national level, ELCON (Electricity Consumers Resource Council, in Washington, DC) represents many large industrial firms, and is a good resource for user-friendly information.

But all the preparation in the world does no good unless your PUC or state legislature acts on this issue. Experience has shown only intervention in the process can move that process in the right direction. While the better marketers are already involved (and those that aren't don't deserve your business), customer input is essential to ensure acceptable results.

Waiting on the sidelines for "the other guy," or the PUC, to release you from your utility's grip will only prolong the present situation. Trusting your utility to do the right thing (by reducing its profit margins, selling off its assets, cutting its staff and perks) is naive: no industry has ever done so without the push of competition. Watching others bear the cost of interventions, while you reap the benefits, might give you a free ride on others' success, but experience shows that utilities use that apathy by dragging out proceedings long enough to exhaust opponents' financial resources.

You can help make the right changes happen by supporting interventions into the regulatory process. When energy users financially sustain such actions (directly, through a group, or with an energy partner), the contribution needed from each is small compared to the value of quickening competition: the payback period of such efforts is typically measured in *weeks*, not years. Participating in such efforts will also help you grasp the opportunities to come. Those who do will reap the benefits of that knowledge, for both their facilities and their careers!

APPENDIX 1 - free magazines focusing on the competitive utility marketplace (request subscription card)

Energy Buyer
Christine Strobel, editor
Infocast, Inc.
13715 Burbank Blvd.
Sherman Oaks CA 91401
ph: 818-902-5400
fx: 818-902-5401

MegaWatt Markets
Randy Rischard, managing editor
Pasha Publications
1616 Ft. Meyer Drive Suite 1000
Arlington VA 22209
ph: 703-816-8626
fx: 703-528-4296

PowerValue
Greg Porter, publisher
Intertec International Inc.
2472 Eastman Avenue, Bldg. 33
Ventura CA 93003
ph: 805-650-7070
fx: 805-650-7054

APPENDIX 2 - Web sites addressing deregulation issues

Strategic Energy Ltd. (best site for state updates)
http://www.sel.com

Direct Access Working Group Workshops (California deregissues)
http://162.15.5.2/wk-group/dai/

Welcome to Convergence Research (general electric industry)
http://www.converger.com/

NYMEX Electricity Financial Tools (helps to understand futures
and other instruments)
http://www.nymex.com/contract/electric/intro.html

The MCGI Home Page (good links and other data)
http://www.mcgi.com/

Electric Restructuring in California (Calif. PUC electric
deregpage)
http://www.cpuc.ca.gov/elec.shtml

Newspage for Retail Wheeling (just what it says)
http://www.newspage.com/
NEWSPAGE/cgi-bin/walk.cgi/
NEWSPAGE/info/d13/d4/d10/

Electric Utility Information (good links and other data)
http://home.ptd.net/~sjrubin/electric.htm

NARUC Home Page (Natl. Assoc. of Regulatory Utility Comm.)
http://www.erols.com/naruc/

The Utility Connection (good links and other data)
http://www.magicnet.net/~metzler/index.html

Energy Central Home Page (headlines and synopses of the day's
energy news)
http://209.31.214.202/EC/MAIN.CFM

The National Council on Competition and the Electric Industry
(group in formation)
http://www.erols.com/naruc/nccei.htm

Utility Deregulation Project (from the Minnesota renewable energy perspective)
http://www. me3.org/projects/dereg/

The Power Marketing Association (day's news and other information)
http://www.powermarketers.com/main.htm

New York State Public Service Commission (just what it says)
http://www.dps.state.ny.us/

misc.industry.utilities.electric Web Site (good links and other data)
http://www.digiserve.com/cpreecs/miue/

The Electric Utility WWW Resource List (good links and other data)
http://sashimi.wwa.com/~merbland/utility/utility.html

Energy OnLine (LCG Consulting Corp., news and other data)
http://www.energyonline.com/

Gridwatch.com Global Power Directory (formerly "Energy Yellow Pages")
http://www.gridwatch.com/

GEM: Global Energy Marketplace (formerly "Virtual Library: Energy")
http://gem.crest.org/

Power Providers (electric service upgrades in West coast areas)
http://www.powerproviders.com/

Electricity OnLine (news and other data)
http://www.electricity-online.com/

PEAR's Electric Intelligence: Insights on Competition (subscription newsletter)
http://www.peartree.com

LEAP Letter (paid newsletter on restructuring)
http://www.spratley. com/leap

New Energy Ventures (NEV) (major aggregator's site)
http://www.newenergy.com

California Energy Institute (publications page) (good technical treatises on dereg)
http://www-path.eecs.berkeley.edu/%7Eucenergy/

Cons. Energy Cncl. Restructuring Forum (dereg from alternate energy advocate's view)
http://www.cecarf.org/restructuring/

ElectricRates Home Page (good source of load profiles and other data)
http://www.electricrates.com/

"Access Energy"—The California Energy Commission (CEC) (technical energy group)
http://www.energy.cat gov/

Automated Power Exchange - We make electricity...(private power exchange)
http://www.energy-exchange.com/

PowerValue Online Magazine - Articles (good free magazine covering competition issues)
http://www.powervalue.com/articles.html

Public Utility Home Page (good links and other data)
http://home.ptd.net/~sjrubin/pubutil.htm

Energy and Environmental News (good links to publications covering dereg, energy issues)
http://www.serve.com/commonpurpose/news.html

PJM OASIS Home Page (see an ISO in action at no charge)
http://oasis1.pjm.com/index.html

Electricrates MLM/Deregulation Forum (chat room for small time power selling schemes)
http://www.electricrates.com/drforum/drboard.htm

Comparison of "Green" Power Products (just what it says)
http://www.edf.org/programs/energy/green_power/c_providers.html

Yahoo Utility News (generic news source)
http://biz.yahoo.com/news/utilities.html

Continental Power Exchange (another private power exchange)
http://www.cpex.com

Talkpower (mostly electric utility distribution discussions, but some good scuttlebutt)
http://www.talkpower.com/

California Competition Network (marketers trying to improve markets methods, rules)
http://www.gcnet.org

FacilitiesNet Deregulation Forum
http://www.facilitiesnet.com/forums/cgi/get/deregulation.html

FacilitiesNet Energy Forum
http://www.facilitiesnet.com/forums/cgi/get/energy/html

UtilityGuide Information Network for Electricity Users
http://www.utilityguide.com/body_index.htm I

ElectricityChoice
http://www.electricitychoice.com/default2.htm

California Independent System Operator (ISO)
http://oasis.caiso.com/iso/isolnk/splashhouses.html

Energyworld - The global business site for electric power and
energy producers
http://www.energyworld.com/

New York Power Pool
http://www.nypowerpool.com/

New Energy Ventures (customer services site)
http://www.nevservice.com/

California Power Exchange (CAPX)
http://www.calpx.com/

A useful bulletin board that discusses these issues can be accessed by logging on to: manager@aesp.org and entering SUBSCRIBE AESP-NET. You will then automatically receive e-mail covering a variety of energy issues, deregulation being one of them.

A bulletin board focusing on developments in California from an "insider's" viewpoint is DAWGNET. DAWG is an acronym for Direct Access Working Group, which consists of people (mostly energy and services vendors) directly involved in making deregulation work from a nuts-and bolts standpoint. Access by sending e-mail to dawg-net@uspi.org and entering SUBSCRIBE in the body copy. See also their web site (second one down on the above list) to update this information should it not be correct at this time.

APPENDIX 3 - alphabetical list of seminar providers

AIC Conferences
50 Broad Street, 19th Fl.
New York, NY 10004
212-952-1899
fx: 212-248-7374
http://www.aic-usa.com
Yalmaz Siddiqui

All Utilities Auditing Co. (a/k/a Electricity Infosource)
3130 So. Harbor Blvd., Ste.
370 Santa Ana, CA 92704
714-432-0100
fx: 714-432-8805
http://www.all-utilities.com
AUACO@aol.com (Richard Strauss)

American Assoc. of Utility Marketing Executives
P.O. Box 8770
Emeryville, CA 94662-8770
510-450-1815
fx: 510-655-7887
barbarap@aaume.com
(Barbara Pereira)
www.aaume.com

American Business Symposiums
60 Webster Road, Suite 300
Weston, MA 02193
617-736-0800
fx: 617-736-0844

Association of Energy Engineers
4025 Pleasantdale Rd. Suite 420

Atlanta, GA 30340
770-447-5083 X223
fx: 770-381-9865
www.aeecenter.org

Camber Corporation (DOE contractor)
601 13th St., NW, Suite 350
North Washington, DC 20005
202-737-1911
fx: 202-628-8498

Center for Business Intelligence
70 Blanchard Road, Suite 4800
Burlington, MA 01803
800-767-9499
fx: 617-270-6216
registrar@cbinet.com

Chartwell, Inc.
1900 Emery Street, Suite 332
Atlanta, GA 30318
800-432-5879
fx: 404-352-8016
utilityinfo@chartwellinc.com

Clemson University
Office of Professional Development
P.O. Box 912
Clemson, SC 29633-0912
864-656-2200
fx: 864-656-0938
Amy Wright

Economics Resource Group
1 Mifflin Place

Cambridge, MA 02138
617-491-4900
fx: 617-576-3514

Electric Consumers Resource Council (ELCON)
1333 H St., NW, The West Tower, 8th floor
Washington, DC 20005
202-682-1390
fx: 202-289-6370
John Anderson

Enerdata Ltd.
Suite 304, 100 Allstate Pkwy.
Markham, Ontario, CANADA L3R 6H3
905-479-2515
fx: 905-470-0117
www.enerdata.com

Energy Expo, Inc.
5 Lewis Lane
Chester, NJ 07930
908-879-8351
fx. 908-879-8371
www.energyexpo.org

Energy News Data
117 Mercer Street
Seattle, WA 98119
206-285-4848
fx: 206-281-8035
newsdata@newsdata.com
www.newsdata.com/enernet

Energy Institute (Energy Seminars, Inc.)
2001 Holcombe Blvd., Suite 806

Houston, TX 77030-4214
888-353-7451
fx: 713-797-0144
nrginst@aol.com
Joshua Schwager (202-986-6746)
www.obnm.com/theenergyinstitute

Energy User News (Chilton Company)
Mike Randazzo, managing editor
201 King of Prussia Road
Radnor, PA 19089
610-964-4223
fx: 610-964-4647
mrandazz@chilton.net
www.energyusernews.com

E-Source
1033 Walnut Street
Boulder, CO 80302-5114
303-440-8500
fx: 303-440-8502
ndoty@esource.com (Nancy Doty)
www.esource.com

Exnet
c/o The Management Exchange
123 East 54 St., Suite 4C
New York, NY 10022
212-371 -8320
fx: 212-371 -8325
exnet@erols.com
www.exnet.net

GDS Associates
Suite 720, 1850 Parkway Place

Marietta, GA 30067
770-425-8100
Betty Reiber

Infocast
13715 Burbank Blvd.
Sherman Oaks, CA 91401
818-902-5400 X22
fx: 818-902-5401
103116.625@compuserve.com (Jim Naphas)

Insight Information Inc.
55 University Ave., Suite 1700
Toronto, Ontario M5J 2V6
CANADA
416-777-1242
fx: 416-777-1292

Institute for International Research
708 Third Avenue, 2nd fl
New York, NY 10017-4103
800-999-3123
fx: 212-661-6677
us002506@interramp.com (Cheryl Fallick)

International Business Communications
IBC USA Conferences Inc.
225 Turnpike Road
Southborough, MA 01772-1749
508-481-6400
fx: 508-481-7911

International Exposition Company
15 Franklin Street
Westport, CT 06880
203-221-9232
fx: 203-221-9260

International Quality and Productivity Center
150 Clove Road
P.O. Box 401
Little Falls, NJ 07424
800-882-8684, 201-256-0211
fx: 201-256-0205
lmoran@planet.net (Linda Moran) www.iqpc.com/

King Publishing Group
627 National Press Bldg.
Washington, DC 20045
202-662-8565
fx: 202-662-9719
kingpub@access.digex.net

Pasha Publications
13111 Northwest Freeway, Suite 230
Houston, TX 77040
713-460-9200
fx: 713-460-9150
clouser@pasha.com (Gary Clouser)
www.pasha.com

Pennwell Conferences & Exhibitions
3050 Post Oak Blvd. Suite 205
Houston, TX 77056
800-883-8189
fx: 713-690-5674
umbrella@pennwell.com (Beth Baker)
www.pennwell.com

Power Marketing Association
1519 22nd St. S-200
Arlington, VA 22202
703-892-0100

fx: 703-979-4677
keltys@erols.com (Peter Dykhuis)
www.powermarketers.com

Princeton Energy Programme
136-230 Main Street
Princeton Forrestal Village
Princeton, NJ 08540-9759
609-520-9099 ext. 132
www.princetonenergy.com

RER
12520 High Bluff Drive, Suite 220
San Diego, CA 92130-0081
619-481-0081
fx: 619-481-7550
www.rer.com

Strategic Research Institute
500 Fifth Avenue, 11th floor
New York, NY 10110
800-599-4950
fx: 212-302-9850

Univ. of Wisconsin Management Institute
University of Wisconsin-Madison
Grainger Hall, 975 University Ave.
Madison, WI 53706-1323
800-292-8964
fx: 608-265-3357

ZE PowerGroup Inc.
Unit #130 5920 No. 2 Road
Richmond, British Columbia V7C 4R9 CANADA
604-244-1472

fx: 604-244-1675
zelramly@direct.ca (Zak El-Ramly)
www.ze.com/ze

Footnotes
1. *MegaWatt Daily*, June 27, 1997, Pasha Publications, Houston, TX.
2. "Moving to Competitive Utility Markets: Parallels with the British Experience," by Dr. George Backus and Susan Kleeman, in March/April 1997 *PowerValue* magazine, published by Intertec International Inc., Ventura, CA
3. "The Competitive Effects of Transmission Capacity in a Deregulated Electricity Industry," by Severin Borenstein, James Bushnell, and Steven Stoft, published by the University of California Energy Institute, Berkeley, CA, April 1997
4. Planergy, Inc. Web Page, http://www.planergy.com, Austin, TX, February 1997
5. Capstone Turbine Corporation Web page, http://www.capstoneturbine.com, April 1997
6. Product Announcement by SouthWire Corp., Power Daily, McGraw-Hill Publishing Co., New York, NY, April 8,1997
7. Entegrity Packaged Rooftop Air Conditioner product brochure, Entergy Inc., Memphis, TN, November 1996

CHAPTER 6

CURRENT HAPPENINGS IN ELECTRIC UTILITY DEREGULATION

NOW IS THE TIME FOR USERS TO RENEGOTIATE THEIR CONTRACTS

F or energy users, the driving force that makes renegotiating an electric contract realistic is the deregulation of the electric utility industry. Monumental changes are occurring that have the whole industry is chaos at the moment. Utilities are scrambling to retain or gain market share. New alternatives for power supplies will become available. Regulatory agencies are becoming more flexible.

Many users are finding the utilities very willing to change from a rigid approach to a customer oriented attitude in anticipation of further changes in the deregulation process. **Now is the time to renegotiate your electric contracts!**

STATUS OF ELECTRIC DEREGULATION

Deregulation of the electric utility industry began with the Energy Policy Act of 1992. It began at a wholesale level and is rap-

Presented in *Strategic Planning for Energy and the Environment*, Vol. 17, No. 4, by Paul Cunningham, P.E.

idly gaining speed. Since many utility companies' transmission systems cross state lines, wholesale deregulation is managed by the Federal Energy Regulatory Commission. By contrast, retail deregulation (Customer Choice) is regulated at a state level by legislatures and utility commissions.

As a consequence, retail deregulation moves at a different pace in each state. Less progress is usually experienced in states where the utility companies have the greatest influence. It is reasonable to guess that most states will have some level of retail deregulation in the next three to five years.

NEW TECHNOLOGY

Just as in the telecommunications field, new technology is beginning to impact the utility companies' traditional generation patterns. In the 60's and 70's, large central coal or nuclear plants took up to 10 years to build; cost $1,000-$5,000 per kW; and required 10,000 Btu to generate a kW of energy.

Now, gas fired turbines take one year to build; cost $300-$500 per kW; and require 7,000 Btu to generate a kW of energy. These smaller units can be easily located out in the service area, rather than being concentrated in a few central spots. The distributed location of the new units utilizes the transmission system more efficiently.

Better monitoring, controls and communication are allowing the utilities to use the generation more efficiently. Generating units are being more fully utilized.

UTILITY COMPANY REACTIONS

Utility company reactions to deregulation vary. Most are still opposing change. Others are accepting the inevitable and see many opportunities in providing new services in new territories.

All are taking steps to prepare for a future without the security blanket of the regulatory process. There are mergers, downsizing, reinventing, new purchasing alliances, management services between companies and diversifying.

One is reminded of the U.S. auto industry several years ago when everyone realized that they were no longer competitive with the Japanese. Some fought hard for trade barriers to protect them while others got busy and markedly improved their quality and manufacturing proficiency. Now, most of them are quite competitive.

The same will probably hold true with the utility companies. The adaptable will survive. We customers must change also, while taking advantage of their changes so that we can break free from the tightly limited rate structure that we have faced for many decades.

REGULATORY AGENCY ATTITUDES

The state commissioners are currently charged with the responsibility of administering that portion of the deregulation process that is intrastate. As you might guess, with fifty different agencies, there is a wide variation of attitudes. Some say that the degree that they are favorable to the deregulation effort is strongly related to their ties to the utility companies in their states. In some cases, there has been an exchange of personnel that might raise eyebrows.

Commissions in states with high power costs are the most likely to be aggressive in the deregulation process. The movement favoring some degree of retail wheeling is impressive. Key dates are January 1, 1998, when a number of additional retail wheeling experiments began, and the year 2001, when many think there will be full deregulation. In all states, their legislature has a direct effect on the deregulation process. Many states will see a strong legislative push in the next few sessions of the legislature.

INDEPENDENT POWER PRODUCERS

A major change in the way that electric utilities do business has been brought about by the independent power producers (IPPs) who are more agile, cost effective and less hampered by some of the regulatory constraints of the utility companies. Their new units are quicker to install, lower in first cost and lower in operating costs. As a consequence, they are very competitive with the utility companies' older, larger plants.

In spite of a considerable effort to hold them down, they are now adding more generation than the old utilities. IPPs offer significant competition that will be helpful for industrials in the future by driving down power costs and offering alternative sources of power.

POWER MARKETERS AND BROKERS

Another major change is being caused by a new breed of fast-moving operators who have moved over from gas brokering and are now making a market in wholesale power. Power marketers will buy surplus power from one utility and sell it at a profit to a second, who will save money by backing down on some high cost generation or postpone construction of new plants with the purchase. Power brokers do the same but do not take title to the power.

Utilities are selling or retiring their less efficient units and increasing the capacity of their remaining units to meet growing needs. More are also buying power from others rather than generating themselves. As a result, their need of services, such as provided by brokers, is increasing.

Power marketers are doing for the electric utility industry what Sprint and MCI did to the long-distance telephone industry by convincing AT&T customers to change to their service. Some, such as Enron, are now purchasing their own utility companies.

WHEELING OR TRANSPORTING POWER

FERC 888 provides that no utility can charge more for the use of their transmission lines to wheel power than they charge themselves. This prevents the possibility of a major surcharge that would make wheeling uneconomical.

Currently, there are two types of wheeling charges under study. Postage stamp rates, which charge the same regardless of distance the power is moved, and the kW-mile rate which does consider the distance. Transmission service charges are in the $1.00-$1.75 range. It is clear from these figures why there is a rule of thumb stating that it is uneconomical to wheel power more than two utilities away.

COMMODITIZATION OF ELECTRICITY

A very active futures market is another development caused by deregulation. Again, this addition is patterned after the natural gas futures trading. The new market gives utilities a chance to make more money on their surplus generation rather than leaving it idle. They can optimize their generation and power purchasing to lower their costs.

Energy trading and risk management to dampen the effects of the resulting price volatility present options not previously available to optimize the pricing of electricity. Price discovery is also quite helpful in planning future management actions. Hedging strategies, future basis risk, swaps, and related tools made possible by the market at the New York Mercantile Exchange greatly improves flexibility of power utilization.

Gas is cheaper to move than electricity. Electricity is cheaper to move than coal. There is a strong movement to lump all together as just energy in different forms and buying the cheapest form available. For example, a power plant might be located near a coal mine rather than using unit trains to move coal to the power plant. The transmission system would move the electricity rather than the coal.

POWER COSTS

The United States is divided up electrically in regions called reliability councils. The utility companies in each of these councils or power pools, band together to manage power availability within its borders and the inter-ties with other power pools. Although there were power sales on a wholesale basis among the members of each pool, the pricing of this power was not generally known.

FERC mandated that this information be publicly available and we now have access to wholesale power whose price is tracked on an ongoing basis. This is a useful tool in determining what the marginal costs your utility company has to pay for any power purchases. It is interesting to compare this figure with that which your power company is charging you.

STRANDED INVESTMENT

With deregulation, most utilities will have some generation and power purchase contracts which are no longer cost effective due to lower cost power from other sources. Writing off this investment is vehemently opposed by many of them. Their preferred alternative is to pass these costs on to those who exit their system in order to buy power elsewhere. Utility companies are making an extremely strong effort to be compensated for all of this investment that will no longer provide them revenue.

Currently being debated is one mechanism to recover stranded costs called "securitization." It will allow a utility to recover its stranded costs up front in a single lump sum payment. The term means converting into marketable securities ("securitizing") the present value of the revenue stream anticipated to be produced by customer payments of stranded cost recovery surcharges over a period of, say, five to ten years. Under this plan, the legislature or utility commission irrevocably orders that customers must pay a surcharge as part of their electric bill to complete the bailout of the utility with stranded costs.

Since the bonds are likely to be favorably rated, they will bear interest at rates less than the utility's other borrowings, and some of the proceeds of the stranded cost recovery bonds can be used to pay off pre-existing debt and thus lower the utility's overall cost of capital. This may result in a token reduction in electric rates but electric rates would be reduced a greater amount and years earlier, if customers could buy electricity in an open competitive market without paying stranded cost recovery surcharges in the first place.

Many say that there are strong indications that some utilities are inflating their estimate of stranded costs, now estimated at $135 billion. By comparison, the gas industry originally estimated their gas restructuring costs at $44 billion. The final number was around $13 billion. These costs were finally allocated between pipeline companies and gas consumers.

There appears to be little interest among utility companies to follow this pattern. Many prefer to delay retail wheeling until this investment can be depreciated in their normal manner.

UTILITIES' WILLINGNESS TO NEGOTIATE NEW CONTRACTS

For the first time, utility customers are moving from a captive status to having a choice in suppliers. Utility companies can see the day when they must (like private industry) rely on price, good service, customer relations and other benefits to keep their customers. The specter of competition is forcing them to make concessions that were unheard of just a few years ago. Many utilities are attempting to realign their relationships quickly with customers before they lose them to other suppliers.

Chapter 7

The Ten-Step Program to Successful Utility Deregulation for Building Owners and Managers

With the passage of the Energy Policy Act of 1992, the process of deregulating the electric industry was begun. Because of this historic change toward a competitive arena, the utilities, their customers, and energy service providers have begun to reexamine their relationships.

How will building owners, each with varying degrees of sophistication, choose their suppliers of these services? Who will supply them? What will it cost? How will it impact the tenants/ occupants? How will the successful players bring forward the right product to the marketplace to stay profitable? And how will more and better energy purchases, commissioning, O&M, and energy services improve the bottom line?

This chapter reviews the historic relationships between utilities, their customers, and energy service providers, and the tremendous possibilities for doing business in new and different ways. Of particular concern to building owners is who will be best able to supply these services in the future and if they will improve quality and reduce costs.

Presented in *Energy Engineering*, Vol. 95, No. 3, by George R. Owens, P.E. C.E.M.

THE IMPACT OF RETAIL WHEELING

Companies such as the Rouse Company have been making energy supply choices on both new building construction and existing building operations since the first building was built. This has included negotiations with utilities for primary service, business recovery rates, economic development rates, interruptible and other rates advantageous to ourselves and our tenants. We also have participated in the deregulated gas industry and have directly purchased natural gas from marketers and brokers on the open market. With an annual utility bill of more than $85 million, the Rouse Company and its subsidiaries and affiliates are customers of 45 electric utilities. The cost of electricity varies from 3.5 cents to more than 12 cents per kWh. The price for electricity is markedly different in neighboring regions, states, and even across adjacent property lines.

Many believe that electric deregulation will even out this difference and bring down the total average price through competition. Data from other industries support the belief of many that deregulation will result in a reduction in electricity costs from a minimum of 10 percent to 20 percent or greater. A research project report by Mercer Management Consulting, Inc. details the cost experiences of five industries after deregulation. The reduction in cost for those industries has ranged from 35 to 65 percent. The first year reduction ranged from almost no change to a 15 percent reduction. Most utilities are already taking actions to reduce costs. Consolidations, layoffs, and mergers are occurring very quickly. As part of the transition to deregulation, many utilities are requesting and receiving rate freezes and reductions. One utility in the northeast has requested a 25 percent rate reduction for industrial customers and a 10 percent reduction in the large commercial sector.

All of this provides for interesting background and statistics, but what does it mean to those of us interested in providing and procuring utilities, commissioning, O&M (operations and maintenance), and the other energy services required to build and operate buildings effectively? Just as almost every business enterprise has experienced changes in the way that they operate in the '90s, the electric utilities, their customers and the energy service sector must also transform. Only the well-prepared companies will be in a position to take advantage of the opportunities that will present themselves after deregulation. By January 1, 1998, building owners and managers need to be in a position to actively participate in the early opening states. The following questions will have to be answered by each and every company if they are to be prepared:

- Will you participate in the deregulated electric market?

- Is it better to do a national account style supply arrangement or divide the properties by region and/or by building type?

- How will electric deregulation affect our relationships with our tenants?

- Major chains may want to partake in purchasing power on their own.

- Should the analysis and operation of electric deregulation efforts be in-house or by consultants or a combination?

- What are the criteria to use to select the energy suppliers when the future is uncertain?

THE TEN-STEP PROGRAM TO SUCCESSFUL UTILITY DEREGULATION FOR BUILDING OWNERS AND MANAGERS

In order for the building sector to get ready for the new order and answer the questions raised above, I have developed this ten-step program to ease the transition and take advantage of the new opportunities.

Step #1—Know Thyself
- When do you use the power
- Summer vs. winter, night vs. day
- What load can you control/change
- What $$$ goal does your business have
- What is your 24 hr. load profile
- What are your engineering, monitoring, and financial strengths in-house

Step #2—Keep Informed
- Read, read, read—network, network, network
- Professional organizations
- Vendors, consultants, and contractors
- Subscribe to trade publications
- Attend seminars and conferences
- Internet resources—news groups, www, e-mail
- Investigate buyer's groups

Step #3—Talk to Your Utilities (all energy types)
- Customer relations are improving
- Discuss alternate contract terms or other energy services
- Find out are they "for" or "agin" deregulation
- Obtain improved service items (i.e., reliability)
- Tell them your position and what you want, now is not the time to be bashful
- It is possible to renegotiate existing contracts

Step #4—Talk to Your Future Utility(ies)
- See Step #3
- Find out who is actively pursuing your market
- Check the neighborhood, check the region, look nationally
- Develop your future relationships

- Develop ESCO, power marketing, financial, vendor, and other partners for your energy services needs

Step #5—Explore Energy Services Now (Why Wait for Deregulation?)
- "Standard" energy projects such as lighting, HVAC, etc.
- District cooling/heating
- Sell your central plant
- Square foot pricing
- Buy comfort, Btus or GPMs not kWhs
- Outsource your operations and maintenance
- Other work on the customer side of the meter

Step #6—Understand the Risks
- Times will be more complicated in the future
- Length of a contract term in uncertain times
- Do you want immediate reductions now, larger reductions later, or prices tied to some other index
- Do you want a flat price for utilities
- Losing control of your destiny-turning over some of the operational controls of your energy systems
- Some companies will not be around in a few years
- How much risk are you willing to take in order to achieve higher rewards

Step #7—Solicit Proposals
- Meet with the bidders prior to the Request For Proposal (RFP)
- Prepare the RFP for the services you need
- Identify qualified players
- Make commissioning a requirement to achieve the results

Step #8—Evaluate Options
- Enlist the aid of internal resources and outside consultants
- Narrow the playing field and interview the finalists prior to awarding

- Prepare a financial analysis of the results over the life of the project—ROI and Net Present Value
- Remember that the least first cost may or may not be the best value
- Pick someone that has the financial and technical strengths for the long term
- Evaluate financial options such as leasing or shared

Step #9—Negotiate Contracts
- The longer the contract, the more important the escalation clauses due to compounding
- Since you may be losing some control, the contract document is your only protection
- Remember, the supplying of energy is not regulated like the supplying of kWhs are now
- The "Who Struck John" clauses are often the hardest
- Include monitoring and evaluation of results
- How do you get out of the contract and what does it cost

Step #10—Sit Back and Reap the Rewards
- Monitor, measure, and compare
- Don't forget operations & maintenance for the long term
- Keep looking, there are more opportunities out there
- Get off your duff and go to Step #1 for the next round of reductions

IN-HOUSE VS. OUTSOURCING ENERGY SERVICES

The building sector has always used a combination of in-house and outsourced energy services. Many large managers and owners have a talented and capable staff to analyze energy costs, develop capital programs, and operate and maintain the in-place energy systems. Others (particularly the smaller players who cannot justify an in-house staff) have outsourced these functions to a team of consultants, contractors, and utilities. These relationships have evolved recently due to downsizing and returning to the core businesses. In the new era of deregulation, the complexion of how energy services are delivered will evolve further.

Customers and energy services companies are already getting into the utility business of generating and delivering power. Utilities are also getting into the act by going beyond the meter and supplying chilled/hot water, conditioned air, and comfort. In doing so, many utilities are setting up unregulated subsidiaries to provide commissioning, O&M, and many other energy services to customers located within their territory, and nationwide as well.

CONCLUSION

Deregulation of the electric utility industry is here. Competition, customer choice, and generally lower costs for kWhs and end uses of energy will prevail. The electric industry, their customers, and the energy services sector will have to reexamine their methods of operating their businesses. Utilities, customers, and energy service companies will try to get into each other's area of business. Energy efficiency projects will multiply tremendously and the emphasis will move from providing kWhs to providing comfort. Only the building owners and managers that are ready for the new era of electric deregulation will achieve improved comfort and reliability at a lower cost. The "Ten-Step Program to Electric Deregulation for Building Owners and Managers" provides a step-by-step methodology for embracing, conquering, and taking advantage of the new era.

CHAPTER 8

WHAT END USERS SHOULD KNOW ABOUT POWER POOLS

INTRODUCTION

One of the features of the sweeping changes taking place in the United States electricity industry is the development of Independent System Operators (ISOs) as a replacement for centralized power pools. Where power pools do not exist, ISOs are being discussed as a replacement for the present system of individual regional control centers that coordinate operations with their neighbors under coordinating agreements. The advent of ISOs is a significant development and will result in new products being developed and marketed to retail customers by energy marketers. In order to deal directly with an ISO, a buyer (or seller) will have to be a sophisticated operation with the ability to transact electricity on a wholesale level. The price transparency efficiency and wider participation associated with ISOs will lead to the availability of financial instruments that can be used to support flexible and customized pricing products for retail consumers.

WHAT IS AN ISO?

Central coordination of electricity production is a characteristic unique to electricity. The key electricity characteristics that

Presented at Competitive Power Congress by Phillip R. VanHorne, P.E.

require this central coordination are the facts that electricity cannot be stored, it must be manufactured when it is used, and at all times supply and demand must be in balance. No other commodity requires that production of the commodity be coordinated among competitors in order for the entire industry to even exist. The current debates that center on the need for an ISO more correctly should focus on whom will operate the ISO and who will decide the production coordination rules. In order for retail electricity choice to exist, development and implementation of the ISOs is a necessary step that must be completed.

An ISO serves to coordinate the operation of all power plants in a region in order to satisfy the physical operation requirements of the grid. An ISO does not determine the price or the economics of the supply, it provides analysis of the bids by suppliers and determines which plants should operate in a given time frame. The ISO decisions will also consider the operating characteristics of the network so that the final operation meets the reliability requirements of the region. In the Northeast, the ISO's plan is to post Locational Based Marginal Prices (LBMP). LBMP is a spot clearing price that all sellers will be paid and all buyers will pay. Therefore, the bids determine which plants run and the highest bid selected determines the spot price in each area.

While the ISO appears on the surface to be a subject for regulators, utilities and energy marketers to resolve, customers have an even larger stake in the resolution of the issues. First, efficient operation of the electricity system is essential for price minimization and realization of the available efficiencies. Second, without access to the availability of the spot price by all participants, the availability of customized pricing products will be limited to a small number of combinations. Third and most importantly, any delays in implementing the ISOs will delay implementation of open access and customer choice.

WHY DO CUSTOMERS NEED A MARKETER?

Similarly to wholesalers in other industries, energy market-ers will buy the electricity in bulk from a combination of suppli-ers, the ISOs, and combine the physical supply with financial techniques to redeliver the energy to end use retail consumers. The marketers will be required to manage the price volatility as-sociated with the spot market and use their knowledge of the electric transmission system to arrange delivery. Entities such as Plum Street Energy Marketing that have assembled a team of experts in the market area will be positioned to develop custom-ized energy packages that meet the needs and requirements of customers. Products that combine the features of known price component and indexed prices to assure that customers do not pay significant above-market prices may provide the greatest customer benefits.

Full service energy suppliers will have the capability to pro-vide a broad range of energy options to customers including natural gas and electricity. For customers who have the ability to manage their energy sources, multi-fuel energy suppliers will provide supply packages that allow switching between fuels based on supply and economics with the appropriate price pro-tections and options. Behind the meter energy services is another service that when combined with energy commodity supply will assist the customer in minimizing total energy costs—the combi-nation of price and consumption. The energy services option not only looks at minimizing use, but also at total energy manage-ment including level of use, times of use and selection of fuel source. The ability to create an environment where the use of the energy is adapted to react and respond to changing conditions in both the use and supply of energy will represent a significant change in the way that energy use is managed and viewed. The development of electricity ISOs will facilitate that change through the availability of new energy products and services.

WHAT DOES A MARKETER DO?

The existence of full service energy marketing services is a relatively new service. While natural gas marketing companies have been offering transportation gas to large users for approximately 10 years, the era of natural gas, electricity and energy services companies such as Plum Street Enterprises is just emerging. In order to provide a wide range of flexible products, direct interface with the regional ISO will be a requirement. At the most basic level, the marketer will buy, sell, and transact both spot and forward energy supplies. Where appropriate, the electricity purchased either through the ISO or directly from a power plant will be combined with support from specific assets to meet the reliability requirements of the customers. In order to achieve this combination, a thorough understanding of the electric grid and the manner in which it operates will be a key skill that a marketer must possess. Since electricity is a regional commodity that is not readily transferable, the regional knowledge and skill of the marketer is a key requirement to meet the customer needs. Through this combination, the marketer can provide lower customer costs than what they could achieve on their own.

WHAT TYPE OF CUSTOMERS?

Different customer types will benefit from a variety of products and pricing options that will become available in the future. Four different types of customers are outlined here to provide examples of how these customers needs and requirements will be met through the deregulated electricity market.

The first type of customer is the budget conscious whose primary driver is to achieve a known stream of energy costs and value price certainty in order to be able to plan and manage expenses. Institutional customers such as schools and hospitals frequently are driven by the need to have a predictable cost pattern and to avoid surprises as they proceed through their budget year. Electricity products that will fit this type of customer typically begin with a fixed price offering possibly with seasonal or time-of-day discounting depending on the usage patterns. Implementation of regional ISO operations will support the forward market for either options or futures contracts that will allow more competitive price offerings to be developed for customers by enabling the power marketer to utilize risk management techniques to create and manage fixed price exposure. For customers who are exposed to volatility in the price of their product a liquid electricity market will enable marketers to develop pricing options that have a fixed price component combined with an indexed component to keep electricity costs relatively constant as a percentage of total product cost. Again an ISO operation would support the price observability necessary to allow the cross commodity trading required to develop this customer option. A further variation on this concept is to include options on increased volumes at a known price formula.

A second type of customer, distinctly different from the first, are those with a high degree of flexibility and even the option to switch fuel type. These customers include the industrial customers in the chemical, petroleum and gases businesses and also smaller customers with on-site generation options, load control systems. Customers with facilities in geographically dispersed areas that can move production to other sites to capture opportunities in locational price differences also are in this segment. These customers may see the largest change in pricing options associated with ISO operations, since they will have the ability to respond to spot market (as short as 30 minutes) pricing and interruptibility to gain access to the lowest possible pricing. In order to reduce exposure to unexpected price volatility, these customers may also purchase price collars to mitigate spot market risk. Again, the existence of an ISO with information and pricing available to the retail market will result in the availability of these products to the market. Indexed pricing with a tie to the either the customers input or output will be possible for those customers for whom the commodities involved allow cross commodity hedging and risk management. For ISOs such as those in the Northeast that are likely to include locationally differentiated prices, transmission congestion contracts (TCCs) will be a vehicle for power marketers to use to develop pricing opportunities for customers with facilities in multiple grid areas that can also shift production to respond to changes in the pattern of prices across regions.

A third category of customers who will benefit from the pricing options resulting from ISO related pricing, will be those who place a high value on protection against higher than expected spot prices. Somewhat different than the budget sensitive customers these tend to be customers with a product dependent on electricity but without options to quickly switch production locations. These customers include pharmaceutical companies, some paper processing companies and specialty metal processes. These customers will be looking for price capped or fixed prices with deliverability of the product at the specified price guaranteed. The products for this class may include electricity tied to bilateral supply agreements negotiated with a specific supplier outside the ISO structure or supply associated with production assets owned or controlled by the power marketer.

The final category of customer is those less energy sophisticated without professional energy purchasing functions. In order to effectively supply these customers, marketers will be expected to not only supply energy products customized to the requirements, but also provide consulting and advice on how to manage their electricity consumption and purchasing in a deregulated environment. The provider will be expected to evaluate the situation and provide value-added service such as energy audits, financing, load balancing and control to manage energy cost. Products for these customers, many of whom will be smaller in terms of the level of usage, will include discount based rates from a published utility tariff or index or rebate style offerings based on consumption and utilization patterns.

CONCLUSIONS

Each of these broad categories of customers will require different services and will benefit differently from ISO based price products. The more quickly the work shifts from debating the need for an ISO and shifts to implementing the ISO structures that provide the liquidity and price transparency to support these products, the sooner that the products and price options will be available. Without an organized wholesale system of exchange of electricity, true retail access will be very limited. Marketers, such as Plum Street Energy Marketing will utilize the knowledge and skill associated with managing electricity products that include ISO based spot purchases to develop customized packages to meet a wide range of customer requirements in a changing environment.

CHAPTER 9

BUYING POWER IN ANTICIPATION
OF A COMPETITIVE MARKET

T his chapter illustrates how electric power purchase terms, appropriate for the evolving electric environment, were negotiated with different utilities to serve an industrial load at multiple sites. Retail competition between alternate power suppliers is either nonexistent or strictly constrained today in the states where these facilities are located. However, as these states eventually follow the lead of others, this will change in the next several years.

This chapter is intended for those with the task of acquiring power in the emerging environment, particularly those with little or no experience in acquiring power in an environment allowing competitive access to multiple energy suppliers.

It includes what each of the sides in the negotiations held as important; a summary of the key contract features; and a recap of how these features balance the conflicting desires of the parties allowing each side to achieve those goals they value most highly.

THE PROJECT:
THE PURCHASE OF ELECTRICAL SERVICE.

The focus of this chapter, the purchase of electrical service, is one facet of a far more comprehensive project.

Presented at Competitive Power Congress by James L. Menning

Background

The purchaser owns and operates a large diameter, high pressure, high volume pipeline transporting natural gas from the Canadian border to the Midwest. The pipeline is being extended from central Iowa to the Chicago area and expanded to accommodate increased volumes from Canada. To maintain the internal pipeline pressure, which causes the gas to flow, gas compressors are located at appropriate intervals along the pipeline. Until now all of the compressors on this particular pipeline have been powered by natural gas fueled gas turbines. With respect to acquiring electrical service, the project started as a comparison between using electric motors to drive a new compressor at each, or any, of four existing compressor station sites versus powering them with new 20,000 hp gas turbines.

During the regulatory approval process the scope of this extension/expansion project was revised several times. The 20,000 hp gas turbines units were dropped and consideration was given to 35,000 hp units at five sites; a 13,000 hp unit at one location; and a 6,500 hp unit at another.

Scope of Analysis and Negotiations

Discussions were held with the staff of the regional electric reliability organization, representatives of four state utility commissions and over twenty utilities in five states. Contracts were signed to purchase electric power for the two smallest units. In addition, substantial agreement was reached to power two of the larger units with electricity before the decision was made to use gas turbines for reasons unrelated to the negotiations.

PROJECT STRATEGY

Three early decisions were critical in the success of the project:

Consultants

No one in the pipeline's organization had particular experience evaluating and negotiating power purchases of this magnitude and in the emerging environment. Therefore, two consultants were retained to advise on the project. One has particular insight into the emerging environment, including how to frame appropriate provisions to be included in contracts for the purchase of electric energy. The other is particularly familiar with the electric industry in the geographic area where the facilities are located.

Selling the Application

The purchaser realized the need to 'sell' the proposed application to the utilities. To be feasible, the application required a low price. But many utilities are unfamiliar with the characteristics of a load such as this that mitigate the price issue:

Size—the proposed site in South Dakota would have been the second largest single point of consumption in the entire state.

Consistent high load factor—the motors are projected to run at a nearly uniform high load throughout the year. This has a positive effect on the off-peak 'valleys' that are characteristic of typical utility load profiles.

Absence of a large 'in-rush' current when starting. The 'soft start' characteristics eliminate the typical load surge upon motor start.

Mutual Benefit

The contracts had to be 'win-win' situations for the utilities and the pipeline, both now and in the future. Many pipeline organization veterans had experienced the problems gas producers and pipelines had in bringing contracts developed in the previous gas pipeline environment into conformity with the realities that emerged following the introduction of pipeline competition.

THE PURCHASER'S CRITERIA

Four criteria have to be met if electric drives are to power the gas compressors:

Reliability

The electric drives have to be at least as reliable as the gas turbines, measured in terms of the percentage of time they are available to run. Gas pipelines have an obligation to deliver contractual volumes as called upon any and everyday of the year. The availability of the compressors is pivotal in meeting this obligation. Since the fuel for gas turbines is within the pipeline, availability of the compressor is primarily a function of the gas turbine itself. As a mechanical device, an electric motor probably surpasses a gas turbine in reliability. The relevant consideration concerning motor reliability becomes a function of on-site availability of appropriate quantities of electric energy. Comparatively short interruptions in power are acceptable, but extended interruptions are intolerable.

Economical

The electric units have to be economical on an 'all things considered' basis. At least two factors dictate this. First, the company has considerable experience with gas turbines and little experience with electric motor drives of this size. Second, the pipeline's shipper-customers, who ultimately pay for the fuel that is used, have to see that any decision to use electric drives was in their best interests. Many of these shippers are natural gas producers. Using electricity would mean that their product was not being used as fuel to drive the compressors.

Contracts

Any contracts for electrical power purchase need to accommodate the current electric industry environment as well as bridge to the emerging, but not yet fully defined, competitive environment. The term of the contract has to be sufficiently long so that the purchaser has a high confidence that there will be access to a competitive retail market when the contract term expires. Yet if the term is too long he may forego for some years the benefits of the competitive market.

Independence

The analysis and any negotiations had to be carried out on an independent basis. Each of the major equity partners in the pipeline purchaser is an energy company with ancillary operations in the electric merchant business. None of these operations was to have a favored position because of their participation in the analysis or negotiations.

THE REGIONAL PICTURE

The Mid-Continent Area Power Pool (MAPP) is one of the regional members forming the National Electric Reliability Council. It establishes rules intended to assure reliable electric service. A particular rule critical to this project is the percentage of spare, or reserve, generating capacity a utility must provide to cover its firm peak requirements (plus provide a cushion for the potential unavailability of generating units). Other critical rules are the specification of what loads are to be included in 'firm peak requirements' and what comprises 'certifiable' interruptible load. The geographic area covered by MAPP generally conforms to the region defined as 'West North Central' for statistical tabulations. Each of the utilities contacted is a member of MAPP and all of the compressor sites under consideration for electric drives fall within the MAPP geographic area.

The MAPP region is characterized by low electric energy production cost as compared to the rest of country. Many generating units are coal fired, either 'mine mouth' or served by comparatively inexpensive rail hauls. Demand growth has been less than what was estimated when large increments of generating capacity were built. There are transmission constraints in terms of both capacity and distance in moving electricity out of the region. The result is surplus generating capacity and relatively low wholesale market prices.

Retail competition between alternate power suppliers is non-existent or strictly constrained in each of the states within the MAPP region. The seven sites that were eventually considered for electric drive compression are in four of these states. In each of these states the utilities have defined service territories. In two states the utilities have absolute geographic monopolies. In the other two there are potential exemptions from the service territory limitations for new services larger than a prescribed threshold.

THE UTILITIES' (PERCEIVED) POSITION

Through the course of negotiations it became apparent that each utility had its particular set of priorities, biases or uncertainties and reasons for those positions. But there were common elements:

Competitive Retail Access

All of the utilities were adamantly opposed to allowing the purchaser 'retail wheeling' rights. While several acknowledged that some form of competitive retail access was on the horizon, the common theme was 'not now, not for this project.' Paraphrased, the stated reasons for Opposition included: 1) 'It is not legal in this state at this time,' 2) 'To allow any exceptions is precedent setting and therefore to our detriment,' and, 3) 'It directly conflicts with our public/political positions.'

Not verbalized, but evident from the utilities positions, are two underlying reasons. The first is the reluctance to abandon a comparatively comfortable monopoly position. In a competitive environment they would face the potential for lower profits plus costs that may be "stranded" without assurance of recovery. The second relates to the 'level playing field.' Those that believe they are comparatively advantaged fear they will lose the advantage. Those that feel disadvantaged want to delay competition until their disadvantage is removed.

Capacity/Demand Balance

All of the utilities showed a degree of reluctance to take on the obligation to serve a load that may impose a long term on-peak demand, particularly when the service needs to be priced at a level that would be economical for the purchaser. All had available generating capability during off-peak periods and, to varying degrees, most had available on-peak capability. Aside from the price issue, their reluctance is attributed to three types of uncertainty. They see uncertainty in relationship to:

1) how much demand will the utility be called to serve in the future (especially if competitive retail access comes about and the utility gains or loses market share);

2) what will be the utility's <u>cost</u> of providing new peak capacity in the future; and,

3) what will be the <u>market value</u> of peak capacity in the future.

Other than 'low cost power will be demanded by someone' the utilities see few clear answers to these uncertainties.

Incremental Capital Expenditures

At certain sites rather significant capital expenditures are required to provide the service. Since the purchaser is being served from transmission lines, a substation, including transformer and switches, is required. In addition, certain sites required a few miles of radial transmission line in order to connect to the existing transmission grid. The utility approach is to include in the price of the service a component for return on and return of any incremental capital expenditure. Many utilities want to limit their risk by fully recovering the investment over the initial term of the contract. This gives rise to a dilemma. The shorter the term of the contract, the higher the rate component. With a higher rate it is more difficult to justify electric motor drives on an economic basis.

KEY CONTRACT PROVISIONS—WHAT'S INCLUDED

The accommodations the parties made to meet each other's valued needs are intertwined and reflected in four areas of the contract:

Interruptible Service
Service is provided on an interruptible basis conditioned by two major provisions. First, the number of hours in which the purchaser may be interrupted are limited. There is a limit both on the aggregate annual hours of interruption and on the hours of interruption in any twenty-four hour period. Second, during periods when the purchaser would otherwise be interrupted the involved utility will shop the wholesale market for supplemental power. If power is available on the wholesale market and can be delivered to the purchaser, including provision for transmission expenses and line loss, at or below a price prespecified by the purchaser, the utility will acquire the energy on the open market and resell it to the purchaser. Otherwise service is subject to interruption.

Facilities
The purchaser agreed to design, build, own and maintain the substation that is required to receive service. In addition, the purchaser will reimburse the utility for the incremental costs of the radial transmission line, switches and control equipment required to connect the substation to the existing transmission grid. The utility owns and maintains these facilities.

Term of the Contract

Specific provisions of the contracts vary slightly. In general they provide for a bundled energy and delivery service for an initial number of years. Following this initial period there are alternatives. If the purchaser has the right to purchase energy for end-use consumption from an alternate energy supplier the original utility is obligated to provide a delivery service for an additional period of years. Or, if electric service is not economical the purchaser can terminate the contract and use an alternate form of energy to meet his needs. Or finally, if the parties agree the contract can be extended under the previous terms.

Price

The specific pricing provisions of the contracts also vary slightly. They provide for an initial fixed price. The price then escalates for a time at a fixed escalation rate. Finally, as the end of the initial term approaches, adjustments in the price are indexed to certain costs that the utility experiences.

KEY CONTRACT PROVISIONS—HOW
CRITICAL NEEDS WERE MET

Meeting a specific need may be the result of several provisions rather than one. But to avoid excessive detail the needs that were the principle beneficiaries of certain sections are shown below.

Interruptible Service

With the limiting provisions attached to the interruptible service the purchaser attained an acceptable degree of reliability. For the purchaser the 'buy through' provision turns the concern related to interruptions into an economic question rather than a question regarding a local utility's available generating capacity. The purchaser ascribes to the concept that in a competitive wholesale market energy will be available at nearly all times at some price. The question is whether the service is worth the price.

The utilities gain the right to serve a load which helps fill their off-peak demand valley. During the primary term of the contract they plan to manage their demand peaks within the limitations on interruptions and avoid an obligation for additional peak period generating capacity.

For several utilities, with whom deals were not done, the 'buy through' provision was a form of the dreaded 'retail wheeling.' The successful utilities point to differences: They, the utility, is the party that shops the market and they, the utility, buys and takes title to the energy. For them it is simply an extension of their energy buying activity, except in this case it is resold at a price which has a direct relationship to the cost of an individual package of energy.

Facilities

In the state where service contracts were finally signed, the rules on authorized service territories are very rigid. In each of these two cases the purchaser moved the location of the compressor stations about a mile from the original proposed site. This allowed the location to be served by a different utility more amenable to meeting the needs of the purchaser.

Since the purchaser intends to make use of these facilities far longer than the relatively short initial term of the contract, they are not faced with the utilities' imperative of recovering the cost over a comparatively short time period.

Title to the transmission facilities remains with the utilities due to their capability of providing maintenance and because of state regulatory restrictions on who may own transmission lines.

Term of the Contract

For the purchaser the term of the contract, with its initial fixed period, gives a reasonable assurance that:

- the price will be reasonable during the fixed period.

- competitive retail access will be available by the end of the fixed period. However, it will probably not be available for such a long time that it represents a significant lost opportunity due to having been locked into a contract with an extended term.

- if competitive access is available at the end of the fixed period, the purchaser has assurance of a reasonably priced delivery service required to deliver the energy to the site.

For the utility, despite the aforementioned uncertainties, the contract term dovetails nicely with their best projection of when the growth in market demand will require them to add generating capacity.

Price

For the purchaser the price meets the requirements of being economical on an 'all in' basis when the cost of electric drives are compared to the cost of gas turbine drives. The supporting computational analysis included the difference in the initial capital expenditures, including the substation and transmission line; the difference in annual maintenance costs and the difference in cost of the actual energy.

For the utility the price provides, perhaps, a lower unit margin than their customary sale. However, the size and comparative assurance of the load, the off-peak benefits, and the fact that the term is short enough that they have a reasonable estimate of costs make the deal worth doing.

CONCLUDING: COMMENT

Purchasing electric service in an evolving environment is a challenging task. Two factors can be of great assistance: Understanding both the utility's and your own company's needs, biases and uncertainties is one. The second is to creatively work towards mutually satisfactory solutions, including those outside traditional norms. Unfortunately many utilities haven't learned to market in a competitive environment. Therefore more of the "marketing" is forced on the customer if the desired results are to be obtained. In particular, this involves the customer learning more about the utility and "selling" his situation to the utility.

CHAPTER *10*

LESSONS LEARNED FROM REQUESTS FOR COMPETITIVE POWER PRICES

The market for electricity is gradually opening for competition. This process is slow and currently proceeding on a state-by-state basis. Some states are allowing retail access to competitive power supplies for all customers. Others are limiting participation to selected consumers in so called pilot programs or in phases as direct access is introduced over a period of several years. At present, competition is restricted to power as a commodity. A few states allow customers and/or their suppliers to provide metering and billing services at competitive rates. No states allow customers direct access to wholesale power markets. This contrasts with deregulation of natural gas, where large, so called non-core or transportation customers can purchase directly from wholesale markets. Experience by large, retail customers with natural gas purchasing has lead many to believe that lessons learned in that market are directly transferable to the retail electricity market. That is not necessarily the case, especially for government agencies who typically have retail accounts that vary widely in size, location, and service requirements. This chapter is an effort to extract early lessons learned by government agencies from their efforts to procure power in competitive markets. It is hoped that these lessons will not have to be relearned by each agency in each state as deregulation proceeds.

Presented at Globalcon '98 by Michael Warwick

ON THE JOB LEARNING

The lessons contained in this chapter were developed by monitoring the competitive power procurements of a hand-full of Federal and other government agencies. This process included collecting and summarizing the initial request for proposals (RFP) followed by interviews with selected procurement staff at each agency. Of greatest interest were remarks about what was expected from the RFP process and what resulted. In almost all cases, initial RFPs were developed through the use of consultants. In all cases, retail direct access was not yet available. Consequently, these consultants had knowledge of competitive gas markets or wholesale electric markets, but no direct knowledge of competitive retail electricity markets. In other words, the consultants and their clients were equally ignorant about what to expect. Nevertheless, there is security in numbers and the consultants were uniformly valued for their insights into the motives of power sellers and the workings and jargon of electricity markets. They were especially valued when market reactions to RFPs did not meet expectations.

In general, the objective of government RFPs, in priority order, are:

- Monetary savings, generally due to current or pending budget cuts,

- Responding to requirements that all goods and services must be competed if competition is available,

- Consolidating billing for all electrical accounts with a single supplier,

- Improving the agency's ability to manage energy use through improved metering and data, and

- Integrating energy management options with power procurement.

Similarly, the apparent objectives of marketers responding to government RFPs, again in priority order, include:

- Building a strategic relationship with a key customer,

- Selling value-added services (audits and efficiency services mostly), and

- Providing electricity at bargain prices.

It should be obvious by comparing the two lists that a significant gulf exists between the two sets of objectives. This led to often significant problems clearly communicating between the agency and vendors and generally resulted in numerous amendments to RFPs and a substantial effort clarifying RFPs to vendors, usually on the phone. Typically, both buyers and vendors used a team approach to the procurement. Vendors often relied on team members that were not all located in one place, requiring complex conference calls to conduct simultaneous conversations. If communications were not simultaneous, teams (on both sides) had to spend time during the calls covering old ground to reestablish where in the negotiation process they were. These problems are not unique to electricity procurement and generally result from government agencies trying to do "too much" in a single RFP. This may be easier to accomplish once retail competitive environments are better understood (e.g., if RFPs are issued after retail competition starts rather than before) and once the market matures.

The number of lessons that can be learned from this process is almost unlimited due to the variations in how each agency approaches the market and how each state deregulates. However, the following were determined to be the lessons with the greatest impact. It is expected that future versions of this chapter will expand on this list.

1. You Can't Get It Wholesale
2. Uncertain Loads and Unclear Prices Go Together

3.   Load Shape Matters
4.   Bigger Isn't Better
5.   A "Billing Agent" and a Contractor Aren't the Same
6.   Price Discounts and Bill Reductions Aren't Comparable
7.   Dollar Savings are Hard to Capture
8.   Rate Discounts (Standard Offers) Are Competition
9.   It's Called "Best and Finals" for a Reason

1—YOU CAN'T GET IT WHOLESALE

Wholesale power markets are active across the nation. Typically, power trades in these markets at an average of 1.5 to 2.5 cents per kilowatt-hour (kWh). Prices in these markets are easy to obtain. Unfortunately, this has led some buyers to assume that these prices are available to retail electricity purchasers. This assumption would be reasonable if gas and electricity markets were deregulated the same way. However, this is not the case for reasons too numerous to explain here. Similarly, retail customers with many small accounts look with envy upon the low prices granted to large customers with single service addresses. This has led to an assumption that sheer quantity of power purchases should result in similar treatment. These assumptions are inaccurate.

There are three primary reasons why these assumptions are not being realized in retail electricity markets.

First, gas deregulation is governed by Federal law, administered by the Federal Energy Regulatory Commission (FERC). Although FERC has deregulated wholesale electricity markets on terms comparable to those it followed with natural gas, it does not have authority to extend comparable terms to retail electricity customers. Instead, states have authority over retail electricity transactions, including those for very large customers. No state is proposing to allow retail electricity customers direct access to wholesale markets. One can only assume that states do not believe retail electricity marketers and consumers are sufficiently sophisticated to handle the responsibility associated with direct access to wholesale markets (this would require the ability to instantly interrupt service to individual customers). As a consequence, state deregulation rules require electricity consumers to continue to purchase bundled delivery service. These services vary by state based on how direct access is implemented (e.g., power pools versus bi-lateral contracts). In essence, power purchases come bundled with required delivery services including reliability requirements that may not be needed or desired by the customer. This adds costs to the wholesale price, complicating direct comparison of wholesale prices and retail price bids.

Second, retail delivery costs, even for gas, vary based on size. Size is determined by consumption at a specific meter. This establishes the utility account that is, in turn, used to associate the customer/account with others of similar size, a customer class. Utilities allocate their service costs to customers based on these classes. In general, smaller customers require more retail service facilities to serve than larger ones. Consequently, they are charged more per unit of energy delivered. As a result, a simple summation of energy use is not an appropriate metric for determining retail service charges (or savings potential).

Finally, deregulation at the state level is not immediate. Instead, it is a transition from a price regulated market for electricity to one where electricity prices (and some delivery services) are set in competitive markets. Some services will remain regulated by the FERC and states. In order to ease the transition to free markets, states have allowed certain expenses to continue to be included in utility bills. These include surcharges for utility stranded costs (revenues lost to competitive power suppliers), continuation of energy efficiency and renewable generation program funding, and taxes for social service programs for low-income customers and local governments.

The lesson here is that it is unreasonable to expect wholesale electricity prices to translate directly into retail price bids. Instead, retail prices will include additional charges that include delivery of wholesale power to retail markets and surcharges and taxes allowed by state regulation that are not reflected in wholesale markets. Thus, price quotes tied to wholesale prices are unlikely to be so-called index minus prices. Instead, they are likely to be "wholesale price, plus," bids.

2—UNCERTAIN LOADS AND HIGH BIDS GO TOGETHER

A common strategy among government agencies is to aggregate the needs of many meters from agencies and facilities scattered around a region. One of the problems with so-called aggregation is that the responsibility for the bills of the loads that are being aggregated is not also aggregated. As a result, the agency acting as an aggregator has to negotiate as an agent for the facilities it hopes to buy for without the ability to actually commit the loads they represent. This requires a two stage negotiation process that is fraught with risk for all parties. First, the aggregator must solicit potential loads to aggregate. In order to maximize participation and thus, the total size of load, they generally allow participants the option to opt out of any resulting deal if it is not in their interest. The second step then, is to negotiate for the best price possible to retain participation. However, marketers are faced with committing to a specific price without a reciprocal commitment by the aggregator to specific loads (including total energy requirements and overall load shape). Finally, once the price bid is secured, the aggregator (or sometimes the winning vendor) must renegotiate with participants to obtain load commitments.

Ideally, customers approach the market with specific service requests in terms of total load and load shape. This ideal is rarely realized because of normal variations in loads due to changing missions and climate. So, marketers always expect to take some risk about actual volumes and load shapes. However the uncertainty presented by uncommitted load aggregations is orders of magnitude riskier. As a result, best price offers rarely result. Another difficulty with this kind of RFP is that marketers are uncomfortable with providing aggregators with a price bid that is widely shared with participants prior to making a formal load commitment. Their concern is that this provides participants with the opportunity to compare competing offers "on the side." This may expose the RFP vendor to unfair competition as well as making potentially proprietary price information public.

The lesson here is that the RFP process tends to be viewed by agencies acting as aggregators as a license to shop and by vendors as a confidential negotiation.

Participation in RFPs by vendors is expensive. It takes a substantial commitment of resources that could be used to pursue other deals as well as costing considerable amounts of money. Agencies acting as aggregators have an obligation to make explicit exactly what they are offering to buy and what is contingent and structure their solicitations accordingly. They also have an obligation to keep pricing confidential when so requested. This puts government aggregators in an unenviable role.

3—LOAD SHAPE MATTERS

Most vendors trade in wholesale power markets. Prices in wholesale markets vary by time-of-day and season. Large vendors trade in multiple markets. This allows them to optimize their purchases to take advantage of price differences between markets. Vendors must be able to match the customers' patterns of use to expected price fluctuations in wholesale markets in order to pass along the best price. Consequently, they need to make assumptions about the volume and shape of electrical use. Obviously, these assumptions are not perfect and each vendor assumes some risk or else lays that risk in the lap of the buyer.

As a starting point, the buyer is well advised to provide potential vendors with the best available data for making these assumptions. Typically, this includes providing:

- Historic energy use data in the form of monthly utility bills for all accounts,

- Hourly load shape data from time-of-use meters installed on very large accounts, or similar data provided by time-of-use meters installed by the customer,

- Operating schedules (in the absence of time-of-use data),

- Expected major changes to historic operations (new uses, closure of facilities, energy conservation actions, etc.), and

- Access to electricity production equipment, fuel switching capabilities, load management systems, and energy storage facilities.

It is widely assumed that loads that are flat (do not vary on a daily or seasonal basis) will attract the lowest price bids. Evidence from recent RFPs is mixed on how load shape affects price. Some vendors are better able to serve loads that vary. However, there is ample evidence that the absence of load shape information results in less attractive price bids. This presents two challenges for buyers. First, some utilities have instituted a practice of charging to provide this data. In general, billing data is the property of the customer, so such charges should be reasonable. However, some utilities are charging as much as $1,000 per account for this data.

These charges are clearly unreasonable and should be protested before local regulators. The second challenge is how far to go in collecting this data. There are obvious diminishing returns. In general, a few large loads dominate the volume and shape of loads. Accordingly, buyers should concentrate on providing load data for these loads. In terms of quantity of data, time-of-use data are often collected on a 5- or 15-minute basis. Generally, this level of detail is not necessary. Hourly data should be sufficient. Also, given that history is, at best, only a guide to the future, 12 to 18 months of data should be adequate. Data should be provided in its raw, 5- or 15-minute interval format if the local utility charges extra for reducing the data to hourly or other frequencies.

The lesson to be learned here is that load shape does matter. Virtually all RFPs include monthly energy use data at a minimum. Larger, more sophisticated RFPs also include load shape data. The challenge to both vendors and buyers is how to provide what is often a complex mass of numbers. One method is to post this information on a public web site. This is a good model to follow for reasons that will be discussed next.

4—BIGGER ISN'T BETTER

Government agencies prefer to negotiate with a single supplier for all of their energy needs for a variety of reasons. Many commercial chains pursue the same strategy. However, governmental energy requirements typically exhibit considerable variation in the size and location of accounts whereas commercial chains have more uniform loads and central locations. As a result, aggregation may not lead to the discounts expected. A good example is provided by the Postal Service. The United States Postal Service typically operates at least one sorting facility and numerous neighborhood post offices in most urban locations. Sorting facilities are 24 hour a day operations and are large electricity customers. Neighborhood post offices are substantially smaller in size and energy demand. They are obviously scattered fairly uniformly across and the urban and suburban landscape. Local post offices are also an eight-to-five type of operation that is typically vacant much of the time. So, although a vendor may be attracted to the load represented by a sorting facility, they are not as interested in local post offices, especially when they may be spread across multiple local utility service areas.

The goal of a single provider presents a trade-off between convenience and lowest price. This cost of this trade off is not uniform among all suppliers. In other words, some vendors may be more willing and able to act as a single supplier for a low price than others. Therefore, it is reasonable to request quotations to serve all accounts, but vendors should be allowed the flexibility to pick and choose. The worst case would be that a vendor would propose to "skim the cream" and offer a low price for the large, most profitable loads and leave the remainder to be served by the default supplier (normally the current local utility). Nevertheless, they may still offer an attractive price that makes this worthwhile. Regardless, such an offer would be superior to the current supplier situation where all accounts are served by different local utilities and no discounts are available. Typically, the vendor will include in their offer an option to consolidate data collection, even from accounts they do not serve. They may also offer to act as a billing agent for these accounts providing the appearance of a single supplier.

The lesson from this example is that RFPs should be clear in the importance of both price and overall bill reduction so that potential vendors can identify their best strategy for consolidating loads to serve directly and which to serve indirectly. One way to do that is to make load shape data for large accounts readily available, such as on the web as suggested above. This will allow potential vendors to mix and match various accounts to evaluate how they can offer the best price, service, or combination of price and service.

5—A "BILLING AGENT" AND A CONTRACTOR AREN'T THE SAME

Normally, marketers serve loads directly using power they supply. Alternatively, they can act as a billing agent for a customer paying the bills for service from another supplier, usually a default supplier offering an impossible to beat price discount. This situation comes about when the transition to competition allows alternative suppliers but the rules limit the savings available in order to protect incumbent utilities. It also occurs where publicly owned utilities, such as municipal utilities, are outside state regulation. When an agency requests bids to serve accounts in such an area vendors have to make special arrangements to include accounts in regulated service territories in their bid.

Unfortunately, both deregulation legislation and government procurement regulations treat a billing agent differently than an energy supplier. Some deregulation rules prohibit billing agents. Others limit rate freeze protections and rate discounts to customers who remain with the local utility. Designating a billing agent will jeopardize these deals. From a procurement perspective, procuring a billing agent requires different processes than buying energy as a commodity. Accordingly, using an energy procurement may not be a valid method for procuring billing agent bids. Further, billing agents are not in a position to offer any discounts off posted utility rates, therefor expected bill savings cannot be realized.

The situation becomes quite complex when the transition to competition allows competitive market savings over the course of the proposed RFP period but not at the outset. In that case, vendors are either forced into one of two awkward situations. The first is to play the role of billing agents in the early years and then switch to energy providers, which may not be allowed in the RFP. The other is to become an energy provider at the outset of competition and sustain a loss in the early years. Obviously, vendors would expect to make up these losses in the later years of the contract. Equally obvious, this strategy would mean a vendor would be unable to offer significant price discounts when compared to receiving service on a discounted rate until such time as competition results in real price decreases. In other words, a buyer would be better off to take the utility rate discount and hold off pursuing a competitive purchase until savings were available. Unfortunately, this situation may be the rule rather than the exception in most states, and that is the lesson.

6—PRICE DISCOUNTS AND
BILL REDUCTIONS AREN'T COMPARABLE

Wholesale power markets are very efficient. In other words, prices are close to marginal production costs and profit margins are very low. As a result it is difficult to make large profits from power sales alone. Accordingly, a number of energy suppliers intend to make their profits from selling value added services that include energy audits, energy efficiency measures, and operations and maintenance (O&M) services. Many governmental agencies are interested in these services as a means to reduce energy costs. Ideally, both goals could be achieved through a process that results in bill reductions that could be compared on a head-to-head basis with pure energy price discounts. Unfortunately, that is currently too difficult to achieve.

First, most energy procurements have to use a simple basis to evaluate metric for comparing bids. Typically, that is price. A bill reduction can be made equivalent to a price quote if the buyer (with the agreement and cooperation of the seller) can assume a level of consumption that can be used to compare a pure low price bid to an estimated bill that reduces consumption by an amount at least sufficient to offset the higher costs associated with the efficiency measures necessary to achieve the energy savings. In order to award a contract on this basis the vendor would have to guarantee the energy savings as the bill savings associated with lower energy prices are guaranteed, de facto.

Second, energy savings estimates are prospective. Generally, an offer of energy savings in a competitive energy procurement would have to be made in the absence of the necessary data upon which to base them. One way to facilitate this option would be to provide potential vendors with audit data or perhaps typical projects which bidders could use to provide specific energy saving and cost bids. Regardless, a buyer would still need some kind of guarantee in order to realize savings equivalent to those available from a price discount.

Third, bidders offering offsetting energy savings would need guarantees from the buyer to ensure that they could implement energy savings projects rapidly enough to produce the necessary energy savings and capture these within the term of the contract. This presents two challenges for governmental aggregators. To start, as an aggregator they are generally in no position to provide guarantees of either access or even specific customers to serve. Also, most facility managers are reluctant to give an aggregator, or their contractor, control over which energy efficiency projects to implement.

Finally, even if it was possible to justify an award on this basis, monitoring the contract would present considerable headaches.

Despite the apparent attractiveness of this approach in concept, it is probably unworkable for most government agencies. Nevertheless, energy procurement contracts should include the opportunity to pursue energy savings opportunities with the energy vendor selected. This would allow another avenue to pursue such projects and may provide mechanisms for capturing budget savings from energy price reductions. (See next lesson)

7—DOLLAR SAVINGS ARE HARD TO CAPTURE

Governmental energy charges are paid from line item budgets. Overruns are recovered by cutting programmatic budgets. Unfortunately, savings are generally returned to the government, not the agency. This presents two problems. First, the agency lacks any incentive to take risks that may result in periodic cost overruns but yield overall savings. Second, in volatile commodity markets annual price fluctuations may result in periodic cost overruns as a matter of course. The challenge for procurement staff and vendors alike is to identify a mechanism that allows the agency to capture monetary savings or for the vendor to carry over savings from good years to cover losses in poor ones.

Returning savings to an agency is tricky because a check to the agency would go directly to a general fund account and not to the agency itself. As a result, bids that are based on discounts against payments to a utility (as with billing agents) or an index or option price (as with a price cap option or so-called index minus price) don't work. Equally problematic is accounting for retained savings by the vendor, as those wouldn't show up as actual out-of-pocket savings since they would be retained by the vendor. Potential options may include equalized payments over the term of the contract and credits for value-added services. The feasibility of these approaches has yet to be determined.

8—RATE DISCOUNTS AND
STANDARD OFFERS ARE COMPETITION

Federal agencies are required to compete for goods and services when competition is available. The transition to competitive electricity markets raises questions about when competition really exists. This is a particular problem when the transition includes rate freezes and standard offers that offer discounts off past tariffs but may not allow a large enough margin for competitive offers from vendors other than the incumbent utility. It can be argued that competition exists when an alternative exists to the old, regulated rate, even if it is a state sanctioned rate. Obviously, if there are no competitive suppliers and thus, no effective competition, procurement from the new, lower rate is easy to justify. However some energy marketers are so eager to make the case for competition that they are making offers that appear to be costing them money. This may present a problem for an agency that solicits price quotes but only receives one bid below the standard offer rate. Normally a quote from a single vendor would be insufficient to justify a contract award. However, if the offer is legitimate and provides real savings to the government it can be argued that the quote is competing with the standard offer or frozen or discounted utility rate. This line of reasoning has not been tested, but is unlikely to result in a protest.

9—IT'S CALLED "BEST AND FINALS" FOR A REASON

This final lesson is aimed at vendors. Government contracting is highly regulated. In general, the process allows a fair amount of information exchange and negotiation between the government and vendors. However the structure for this exchange is at the discretion of the procurement staff. Some allow individual discussions between vendors and procurement staff. Some require all communication to take place in public forums. Once bids have been submitted, a short list of vendors can be selected for direct, one-on-one negotiations. These are privileged conversations. None of the conversations at any stage of the process can unilaterally change the nature of the RFP without requiring an amendment to the RFP. Substantial revisions may result in a new RFP altogether. As a result, if a vendor pushes for specific exceptions that may give them a competitive advantage, they run the risk of the whole process being redone using their suggested approach, but with the participation of other vendors encouraged.

After the government is satisfied that it fully understands the elements of qualified vendor's proposals the process proceeds to best and final offers (BAFOs). Best and finals includes a final offer of terms and conditions by each vendor. When BAFOs have been submitted, the government is no longer free to negotiate. As a result any lack of clarity works against the vendor. Similarly, contingencies may void the offer, disqualifying the vendor. The challenge faced by vendors when a governmental agency acts as an aggregator is that the agency is unable to commit to a specific load or load shape while the vendor has to commit to a specific price. The agency will use this price bid to solicit participation by its client facilities which will ultimately result in load commitment. Obviously, this interactive process presents some uncertainty for the vendor. However, they have no choice but to bear that risk. If they attempt to withhold final pricing until they have load commitment they will void their offer.

CONCLUSIONS

Competitive electricity procurements are new to most governmental agencies. Although they have well developed procedures for procuring goods and services from competitive suppliers they are not familiar with electricity markets. Similarly, most electricity vendors are not familiar with government procurement practices. This poses challenges for all parties as competition in electricity markets begins. Mistakes will be made as all parties struggle with this new opportunity. It is expected that sharing of lessons learned along the way will minimize these mistakes and facilitate both the procurement process and the delivery of the benefits of competition to government agencies.

CHAPTER 11

NON-PRIVATE POWER AGGREGATION: NEW MARKETS FOR PUBLIC-PRIVATE PARTNERSHIPS

T he emergence of privately sponsored customer aggregation firms aimed principally at the private market in response to electric power deregulation was an obvious development, following the pattern set first in natural gas. Aggregators— whether independent, or associated with power marketers or ES-COs—bring with them not only the capability to enhance composite purchase power muscle, but to provide to aggregation customers valuable system planning intermediary services to structure lower overall costs, by, for example, better matching usage requirements.

The emergence of "non-private" aggregators in the public and the not-for-profit sector has not received as much attention, because those fields generally have been viewed as additional markets for the private sector rather than themselves entrepreneurial initiators of change. That viewpoint overlooks a salient fact: deregulation threatens these sectors themselves with privatization unless they respond proactively themselves.

This response is now beginning to emerge—in self defense. It presents the private sector with an opportunity novel in the domestic power arena, but increasingly common in other formerly public service fields: the creation of forms of "public-private part-

Presented in *Cogeneration and Competitive Power Journal*, Vol. 13, No. 2, by Roger D. Feldman

nerships" able to justify their existence to consumer citizens in the marketplace, on both an economic and a social basis, and backed by strong political muscle.

It is useful to recognize—and to analyze the ramifications of the fact—that the non-private sector currently is not unitary. It is comprised of three different types of responses to deregulation:

(1)   bottom up "municipalization," whereby local public jurisdictions seek to seize for their political constituents deregulation economic benefits that may go outside of their communities to the large business sectors;

(2)   "top down" public power and coop efforts to transform themselves and thereby preserve their institutional role in the changing marketplace; and

(3)   "clustering" efforts by institutions historically serving not-for-profits in other fields (and concerned with the competitive viability of their not-for-profit customers in a privatizing society).

By focusing on the common and then the special aspects of each of these non-private submarkets—regulatory and institutional—private entrepreneurs can identify a number of concrete possibilities for useful collaboration. Examples of each type of collaboration now can be identified.

The common concern of each of the non-private constituencies has its generic foundation in the fact that FERC Order No. 888 fundamentally focused on two goals: (1) opening competitive markets on the blanket theory that all would be benefited, both pricewise and in terms of product differentiation and (2) providing electric power firms which had been developed in the pre-deregulation environment with transitional reimbursement for prior investments rendered noncompetitive. There has necessarily been some breakage of other interests affected by this approach, notably in the non-private sector.

As a general matter, as a consequence of utility exit fee requirements, the economic rate reduction benefits of wholesale deregulation by FERC have been primarily limited to existing wholesale customers. Many existing public power firms furthermore have found their prospects for competition with imported power not encouraging and now are facing Federal legislative efforts to circumscribe their ability to compete in power export markets, even where they are competitive.

Bundling of disaggregated electric services and packaging of electric services with related end use consumer, convergent non-power services has proved to be slow in its implementation. ISOs now being established at the regional and state level are homogenizing the delivery of unbundled electric services.

Overall, as a result of utility stranded cost recapture, except to the extent of specially contrived securitization arrangements, the benefits to residential retail customers—certainly at least pending the introduction of vigorous effective real retail competition—remains to be seen.

In response to these common related regulatory developments, different groups in the non-private sector have responded differently.

BOTTOM UP MUNICIPALIZATION

One response to emerging disillusionment by consumers to the perceived limited impact of deregulation has been the emergence of "bottoms up" support for community franchising and energy development—notably in upstate New York and New England. In part, these initiatives are linked to the extent of surviving local franchise in the electric power arena, relative to state-regulated monopoly investor owned utility franchising.

The same legal stratagem which virtually ended dispersed industrial cogeneration relative to franchised investor owned companies effectively cut back local franchise authority as well.

The concern of local franchise deregulation critics is clear: blind efforts to wipe out local community authority and forms of community choice in the name of clearing "market barriers" potentially in the absence of public involvement could open the way to predatory market practices by power marketers and in consumer related fields.

Private developers, like Energy Choice in New York, have begun to collaborate with municipalities seeking to address this issue, encouraging municipal use of powers of condemnation of existing investor owned assets; construction of parallel facilities; annexation and "muni-lite" (provision for a minimal level of municipal control as a grounds for wholesale purchase rights). FERC has directed the utility to provide transmission in two such cases and derived one "muni-like" case. The majority of cases are still in process or on hold. It is an on-going battle which the advent of retail competition may or may not obscure.

In New York, one effort seeking to go forward is relying on legal interpretation of Order No. 888 to the effect that if service from multiple alternative private users was available pre-deregulation, then municipalization is available post-deregulation without requirement of the payment of exit fees. Its proposal is the subject of a local referendum, which, if passed, will entitle the municipality to act pursuant to the State's "quick take" eminent domain law, subject however to FERC action.

The key argument underpinning community energy developer claims of substantial future savings is that for individual consumers, only the energy part of their bills—roughly 25% of the whole—will be subjected by competition to deregulation. The balance of such bills will continue to be paid for basic utility services which will still be performed by the original provider, on a non-competitive basis.

If community energy municipalization becomes a more successful trend over time, other forms of private collaboration with public entrepreneurs besides aggregation would become possible—notably public finance of new facilities and securitization.

MUNICIPAL POWER/COOPERATIVES

Whether that will be the case, however, will be a function, in part, of whether existing municipal utilities and cooperatives, as a class, will continue to be viable players on the electric power scene. There is a growing school of thought that public power could only survive in the past in its non-entrepreneurial form because privately owned utilities were themselves so heavily regulated as not to be competitively as aggressive as they might have been.

In effect, to a significant extent, investor-owned utilities (IOUs) and munis each operated in a monopoly service territory. In particular, in the new era, while public power agencies which essentially are distribution systems are not necessarily threatened, those systems which own high cost generation and those that are diverse, diffusely governed joint action agencies (particularly if not very competitive) face considerable difficulties. There are certainly states like Nebraska (10% publicly owned power) where there are low prices, high reliability and community sensitivity all embodied in the power system, as its legislator's commissioned reports have noted.

But many municipal agencies feel compelled to seek, in effect, types of product differentiation from the investor-owned competitors springing up all around them, if they are to stay in business. One key such public power strategy is load aggregation—of all or some of municipal services—both as a revenue source and a basis to enhance competitiveness of assets.

The Energy Authority created by Jacksonville, MEAG and South Carolina Public Service as a public power independent system operator (ISO) is an example. Some major public power firms have taken this further and have entered the power marketing field. This has been more feasible for those public utilities and coops which have voluntarily undertaken the type of functional restructuring which IOUs have undertaken more extensively. Oglethorpe's arrangements which enabled it to collaborate with power marketers are an example.

Nevertheless, it does not seem prudent, as a general matter, for municipals to assume that they can fight their way out of their potential problems simply by combining the utilization of aggregation techniques with a better service package, unless that package is able to be price competitive. It also remains a problematic issue as well whether bundling of electric services with consumer services such as home security, telecommunications, HVAC and energy services, will—in what is already a privately driven service aggregating marketplace—produce the desired results.

Consequently, it is appropriate for municipal power to be exploring how, through a variety of private sector strategic alliances, it can relate its obvious potential central role as aggregator, to market position strengthening elements which it can gain from the private sector.

These public-private alliances can either be focused on finances, operation (including outsourcing) or diversification. The possibility of reworking the capital structure of municipal power through IOU contract buy down (in effect, the reverse of independent power producer (IPP) stranded cost buy down by IOUs); through use of power marketing; and through long term purchases of future capacity delivery, have only begun to be explored.

While many munis are still thinking they can prepay their existing contracts and hence their way out of their problems by raising rates, this could prove to be a politically unwise strategy. There is room, in short, for public/private strategic alliances related to finance, without reference to power sales contract restructuring and risk management.

The extent of muni financial flexibility may be further enhanced through strategic alliances with the energy marketing firms that complement the munis' core strengths (e.g., customer presence and relationships; distribution/maintenance services) with their own in areas such as the development of innovative customized products and services; multi-commodity and risk management skills, wholesale trading capabilities, and financial flexibility in structuring asset outsourcing. The essential point for munis is that existing assets may have to be used in new ways in the lower cost, more competitive deregulated market. Introduction of skilled private parties not only may provide new ideas in this regard, but new mechanisms to effect restructuring consistent with the special legal requirements governing public power financing, using, for example, derivatives.

Diversification into other services (or exploitation of the franchise available extending even to telecom uses of rights-of-way to provide such services) very likely also can be effected better with a private partner.

Not-for-Profits

The logic impelling government to acquire the benefits of aggregation in order to continue to serve their constituencies, applies as well to non-profits in fields such as health and education. Privatization impinges on them and their service institutions in different ways.

As social services are deregulated, these non-profits stand to be driven from the marketplaces they serve by for-profit entities. Health care, training and welfare/social service administration are concrete examples. The pressure of privatization moves them in the same way the pressure of competition moves private firms to control energy costs. In turn, that pressure moves the institutions which serve the non-public sector.

Institutions servicing the large market which non-profits represent sometimes are in a more flexible position than governments to extract special benefits from private providers of aggregation and other energy services. A good example of the exercise of this potential is the Massachusetts Health and Education Financing Administration ("HEFA"), which branched into aggregation, and sought a panoply of useful services for its members—thus enhancing the value of its aggregation package. These included:

(1) Supply Service—firm, with delivery assured to one or more points of service location; on two and five year contract bases (including time of use, fixed price and indexed offer);

(2) <u>Green portfolio supply option</u>, with commitments to expand the source;

(3) <u>Special tailored enhancements</u> to meet the requirement of individual participants; and

(4) <u>Program for employees as well as for institutions</u>.

The HEFA procurement package provides, in effect, for institutional consumer protection assurance. HEFA's goals include expansion into prepurchase of electricity and expansion into natural gas, heating oil, propane and perhaps even telecommunications.

Non-profits are not generally captive to long term energy purchase use arrangements designed to facilitate financing, like municipals; dependent on the operation of public laws governing procurement and partnering, like municipalizers; or committed to the use of taxable finance, like private aggregators or outsources.

They are well situated to link aggregation to consumer social preferences (e.g. "green" power) or special economic needs (e.g. dispersed energy for high tech applications.)

Overall, then, they possess strategic advantages which make them natural aggregators and partners for private firms.

CONCLUSIONS

In sum, motivated by needs to survive the twin thrusts of deregulation and privatization, the "non-private" sector is beginning to take the initiative in the energy aggregation field. Success in establishing public-private partnerships with it requires appreciation of the legal and also the institutional survival issues with which they are dealing. The results can be creative uses of municipal franchises, a renaissance of non profit power firms, and the continuing financial vitality of the seriously threatened and socially valuable not-for-profit public sector.

From the perspective of private firms, aggregation by non-private firms can be a portal to response to the electric power deregulation revolution.

CHAPTER 12

EXPERIENCE WITH LARGE-SCALE LOAD AGGREGATION AND POWER PURCHASE BIDDING FOR STATE FACILITIES

T he author reviews the background of restructuring of the electrical industry in California, and the experiences of major state agencies with large energy bills in their efforts to take advantage of the opening of competition for electricity generators coming with the opening of direct access to the generation sources by the customer. Included are the process followed, experience in aggregating loads between major agencies each with multiple locations, and the difficulties in obtaining customer-oriented results in regulatory processes which historically have been managed successfully by the large investor owned utility companies.

BACKGROUND ON ELECTRICAL RESTRUCTURING IN CALIFORNIA

The Federal government in 1992 passed the Energy Policy Act of 1992 (EPACT 1992), to encourage various energy conservation measures and allowing competition in the retail electricity

Presented at West Coast Energy Management Congress '98 by Carroll E. Winter, P.E.

market, to be implemented by the individual states.

Following EPACT, the California Public Utilities Commission (CPUC) in April 1994 proposed to restructure the electrical industry in California. The primary reasons were to allow customers to have a choice of suppliers and to force down electric rates, which in California were 50 percent above the national average.

After many hearings at the CPUC, the Commission made a decision to implement the restructuring in December 1995. Then, following marathon debates at the California Legislature, Assembly Bill 1890 was passed unanimously and signed by the Governor in September 1996. This established restructuring as a matter of law, targeting January 1, 1998 as the starting date for supply of competitive power (referred to as direct access).

The existing tariffed rates of the investor owned utilities (IOUs) were frozen by the legislation until the end of a competition transition period on March 31, 2002. A Competition Transition Charge (CTC) was established for all existing and new customers in order to obtain an early write-off by the IOUs of noncompetitive generating plants and existing above-market supply agreements with independent power producers, in order to help make California utilities competitive in rates by the end of the transition period.

It also created a Power Exchange (PX) to establish a market, manage the bidding of electricity supply and demand and oversee the hourly prices; and an Independent System Operator (ISO) to manage the transmission power grid. All IOUs are required to sell their generated power to the PX, and buy power for their customers from the PX. Other generators may also sell power through the PX, or may choose to sell through bilateral agreements with customers. All power is to be transmitted through the ISO transmission grid.

Inasmuch as the CPUC does not have authority over municipally owned utility districts, the municipals (referred to as muni's) are not required to open their territories at this time to competition from outside generator/marketers.

THE CSU/UC PROCUREMENT PROCESS

Having followed the proceedings at the CPUC since April 1994, and their implementation decision of December 1995, the Chancellor's Office of the California State University (CSU) commissioned a study in early 1996 to make recommendations on how to procure energy in the coming new environment. This study looked at how all utilities were purchased and used at the 22 campuses throughout the state.

Following a review of the study in November 1996 by a committee of the CSU Chief Administrative and Business Officers (CABO), a team of consultants under Grueneich Resource Advocates was retained to work with the CABO committee and staff from both the Chancellor's Office and campuses, to prepare detailed planning and documentation for procurement of electricity on behalf of the CSU System.

Shortly after starting work, we became aware that the President's Office of the nine campus University of California had retained the same consulting team for a similar task. After comparing our needs, an agreement was quickly reached to combine our efforts in order to make better use of resources and to enhance our purchasing power in the market

A task force was established to represent the interests of the campuses as well as management of the two university systems. The task force included campus vice presidents, CSU Chancellor's Office energy and procurement specialists, UC President's Office energy and legal specialists, and energy managers from both university systems.

The present electrical needs of the two university systems are as shown:

California State University System with 22 campuses, $40 million/year for electricity, 130 megawatts of demand;

University of California System with 9 campuses, $58 million/year for electricity, 225 megawatts of demand.

The size of our power needs clearly established us as an attractive potential customer. Early in 1997, many national and local energy firms including unregulated subsidiaries of the investor owned utilities were indicating their plans to supply competitive power to California. We decided to do a two-step procurement in order to narrow the field of suppliers to those who were well established, financed and organized to be considered feasible candidates for what was going to be a complicated, expensive undertaking by power marketers.

We advertised for interested firms to supply electrical services, and prepared a substantial Request For Qualifications. This RFQ required information on the company's history, experience of their management team and service providers in the energy field, financial information and credit history, proposed organizational structure for this new market, and degree to which they had dedicated staff to the California market. Specific experience and ability in procurement and marketing of energy and in scheduling electricity delivery on an hourly basis were requested.

Following receipt and review of statements of qualifications in June 1997, nineteen firms were found to pass the threshold requirements. After interviewing, five of these were selected to receive the next level of selection. Based on information reamed during the RFQ process, we spent about two months reviewing data and preparing a Request for Proposal. The RFP emphasized our search for best pricing for the electrical commodity and related services such as metering, billing and reporting, plus offerings for optional energy related services.

It was our intent to receive bids in October, 1997 in order to have a contract fully in place before the January 1, 1998 start of direct access. Ongoing proceedings on the restructuring before the CPUC, even through December 1997, heavily favored the existing utility companies rather than encouraging direct access. As a result, many national and some local firms pulled out of the potential California market, including some of the five firms on our short list.

Proposals were received on December 8, 1997 from only two firms. One offered the best price on the basis of selection outlined in the RFP, and was therefore chosen for negotiations to reach agreement on contract terms and conditions.

As the 1/1/98 start target approached, the ISO/PX management determined that the software necessary to operate the market and control the grid was not yet ready, and a delay of direct access was announced until 3/31/98.

With the help of an expanded multi-campus negotiation team, discussions with the firm are continuing favorably at this time, with improvement being made in pricing discounts to the universities by extending the time to cover the transition period of four years. Inasmuch as there was no history upon which to base what the PX price might be, the parties have reached agreement on establishing a discount from the entire frozen tariffed rate rather than on the volatility of the unknown PX rates.

LESSONS LEARNED

Final details of the terms and conditions are being reviewed as of this writing, with agreement expected well before the delayed start-up of the direct access process. The implementation process to install the necessary power meters and communication links, create billing and reporting formats, have supplier visits to all facilities and hold workshops to teach personnel what can be expected, is proceeding. Although the start of supply service under the new arrangement has not occurred, we can already report on some experience which may be of interest to others around the country as this movement expands to other states. Among these are:

1) The need and/or value of doing a large scale aggregation of loads to obtain more favorable pricing was an early debate within the universities. We believe that our efforts in doing this were re-

warded in the savings we have obtained. We also found an excellent spirit of cooperation between our two institutions and among the many campuses.

2) Even if there were no savings directly associated with the aggregation, we have gained immensely in sharing consulting and staff time resources, and taking advantage of the various skills available in different organizations. We believe the transaction cost savings for the supplier in putting together one contract for all of these facilities are very real as well.

3) Support and strong leadership from top management is essential to maintain a strong coalition among many diverse interests, personalities, and needs such as multiple university campuses. Opportunities to diversify our customer base further to include non-university organizations were refused, and we think wisely, due to too much diversity of interests and contracting procedures among the various entities.

4) There is no substitute for knowledge of the subject when approaching a complex undertaking such as this. Competent consultants experienced in the legal and regulatory arena, facilities staff knowledgeable of how they use energy and what they need to do their job, and contracting officers who can understand the complications of the energy business all were necessary to accomplish this task.

5) The efforts of a highly skilled and experienced team of regulatory and economic consultants, the digesting of reams and reams of proceedings and regulatory decisions, and being a frequent and vocal presence at the Public Utilities Commission, have resulted in significant savings to the universities in spite of the long regulatory history and financial commitment of the established investor owned utilities in trying to control the process of electrical restructuring.

CHAPTER 13

WHAT END USERS SHOULD KNOW ABOUT REAL TIME PRICING

his chapter describes how Real Time Pricing (RTP) can be used effectively by end users to reduce energy costs. It begins with a description of how this type of rate came into existence, then demonstrates how the rate works, how to optimize benefits through effective utilization and includes an example of how an existing industrial end user benefited through RTP.

BACKGROUND

In the old days (say more than 5-10 years ago), end users with a burning desire to reduce their electric costs had a few choices:

- Physically move to a lower-cost utility

- Make major capital investments to improve energy efficiency

- Produce less product

Since all of these choices have somewhat severe drawbacks, they were not particularly popular nor cost effective.

As competition grew between utilities, electric utilities began

---

Presented in World Energy Engineering Congress by Frank J. Richards, P.E.

to offer incentives to end users to move to their territories or to expand within their territory. They also began to compete more with other fuels in an attempt to expand market share. This gave rise to additional choices for the end user, such as simply *threatening* to move to a lower-cost utility to achieve rate concessions, rather than actually moving... this was very cost-effective for end users.

Innovative rates began to surface, including curtailable or interruptible rates, and real time pricing.

GENERATING INTEREST

As most folks know, electricity doesn't store well... once you produce it, you have to use it. You can't even just lose it... if it isn't dissipated somewhere, strange things start happening, like rising frequency and system instability. These aren't desirable. So progressive electric systems (such as the Pennsylvania-New Jersey-Maryland Interconnection (PJM) economically dispatch generators on their system. The lowest cost units are placed in service first, then as the load on the system increases, higher cost units are added to match the generation to the load.

Also as most folks know, most people are generally more active during the day when they are working, playing and eating, than they are at night, when they are sleeping. Since this activity has a positive correlation to electricity consumption, there is more electrical load at various times throughout the day than there is at night.

So it's not surprising that since the lowest cost generation is run first, that's all that's running at night, and higher cost generators are added as the load builds during the day. But traditional electric rates charged to end users are the same whether the electricity is being used in the day or the night.

This leads to the seed of a win-win revelation, and explains some special rate structures: Utilities realized that if they could get customers to use less during high cost times, they wouldn't have to operate their highest cost generators, so they could save money and offer a lower-priced rate (curtailment rate) to the end user. Also, if they could get customers to use more during the low-cost times, they could see a higher margin as well as offer a lower-priced rate (off-peak rate) to end users.

BIRTH OF REAL TIME PRICING

The extension to this logic gives birth to the concept of real time pricing, or RTP. If the utility would charge the end user with a varying hourly rate that followed the actual cost of generation, the end user would want to consume more during low cost times since he would be rewarded by lower production cost, and want to consume less during high cost times to keep from being penalized by high costs. This kind of rate would provide the incentive of both a carrot and a stick.

So utilities created a rate structure that would provide 24 hourly prices for a day. To enable a return for the utility, the anticipated interconnection price for the energy would be bumped up by perhaps 1 or 2¢/kWh. Although this sounds good, it was a bit too simple, because utilities realized that each end user who took advantage of the RTP rate would no longer be supporting his own weight of imbedded costs spelled out by the regular rate tariffs.

ENTER THE CUSTOMER BASE LINE

To assure each end user carried his fair share of costs, the Customer Base Line (CBL) was invented. The end user's previous year's demand and hourly usage (that's right...*hourly* usage) were used to calculate a base billing at the customer's pre-RTP rate. This establishes the "base-line." The base billing for a particular month is calculated, then each hour where actual usage is greater than the base line is multiplied by the price for that hour and added to the base line bill. For each hour where actual usage is less than the base line, that hour's usage times the price for that hour is subtracted from the base line bill.

If the end user consumed exactly what he did the previous year for each hour, his bill would be the same as the base line bill, because the usage for each hour would not be greater nor less than the CBL. Theoretically, this would mean there is no risk to the end user to be on the RTP rate.

THE DOWNSIDE

But this rate structure requires sophisticated metering and communications capabilities, so the utility charges a monthly fee for them ($350 for PA Power & Light). And even if the end user consumes the same amount of energy for the month, but does it by exceeding the CBL sometimes and being under the CBL other times (which is a very likely scenario), his bill will be *higher* than it was before, due to the way the hourly RTP is determined: Quoting the PP&L tariff, RTP "...is an hourly price determined by the Company from its estimated Marginal Operating Cost, Marginal Capacity Cost, Loss Adjustment Factor, and Risk Adjustment Factor." The Risk Adjustment Factor "...is an adder, not to exceed 1.0¢/kWh which provides a margin over costs on incremental sales and compensates the Company for the risk that hourly energy prices, quoted a day in advance, may vary from actual energy costs."

When the hourly usage is greater than the CBL, the risk adjustment factor is added to the marginal price of energy to determine that hour's RTP to be paid by the end user. When the hourly usage is less than the CBL, the risk adjustment factor is *subtracted* from the marginal price of energy to determine that hour's RTP to be credited to the end user. What this means is that if you use more than your CBL in a particular hour, the RTP applied to that increased usage will be the marginal cost (say 3¢/kWh) plus the risk adjustment factor (say 1¢), or 4¢/kWh But if you use less than your CBL in that hour, the RTP applied to that decreased usage will result in a credit of the marginal cost (3¢) *less* the risk adjustment factor (say, 5¢, or 2.5¢/kWh.

Another downside is that most folks don't have a lot of cost-effective control of the timing of their production...even if the hourly cost gets to 50¢/kWh (which has occurred), the costs for disrupting production would far outweigh paying the 50¢/kWh for the high hours.

THE UPSIDE

Some folks do have production that can be interrupted with little disruption to throughput or excessive manpower cost. For example, a batch metal melter may be able to hold off a heat or two when prices get very high, especially since tomorrow's prices are generally provided to the end user by 3 p.m. today, which gives the end user some time to plan a strategy. Since the production can likely be made up by running an extra heat or two later when the prices are at 2¢ instead of 20¢, this results in savings by reducing consumption during high price times as well as by increasing consumption in low cost times.

Even though there can be very high prices for certain hours of the year, overall the RTP averaged for the year may only be 3 to 4¢/kWh. This can be a big upside for a growing company, because even though they may not be able to alter their usage pattern to avoid high price times and capitalize on low price times, their overall cost before expansion was probably 5-6¢/kWh. So their "new" consumption over the CBL will be perhaps 2¢/kWh cheaper than their base load! This has provided some very real savings for end users in this situation, even though the premise of real time pricing is being ignored.

EXAMPLE

An example of using the RTP rate to advantage involved one of my clients in PP&L territory. Historical RTP data was analyzed to determine when high and low prices were likely to occur at the plant. Then an evaluation of load shapes began with an emphasis on how loads might be altered to take advantage of the real-time prices. In PP&L, which peaks in the winter, daily price curves are generally low during spring and fall, as well as holidays and weekends. Peak days in the summer and winter are typically of two types as shown in the following graphs:

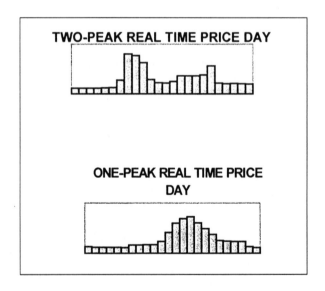

Different operating strategies are required for each type of day. Since prices are very low in the spring, fall, holidays and weekends, no special operating strategy is required, unless production is seasonal and can be shifted to these time periods from peak time periods.

In this client's case, no seasonal variations were apparent. However, daily variations did provide opportunities. Since the electric melting operation only operates two shifts, an opportunity existed to change shift start times to minimize operation during high-price hours in the summer and winter. This was considerably easier to accomplish for the one-peak days, and shifts were arranged to avoid all normal high-cost hours.

For the two-peak days, all the high-cost hours cannot be avoided, but they can be minimized. Depending on the hourly price and production needs for the day, melting can be curtailed by using holding power only. When required, keeping employees 2 hours overtime allows another heat to be run, either to maintain production or take advantage of an extra, low-cost heat.

Making these shift changes resulted in net savings of $9,000/ year. Additional savings were investigated by considering reducing use of some of the pollution-control equipment, when operating conditions did not require its use. The following graph is a simple tool which was quite useful in making operating decisions for running large electrical loads.

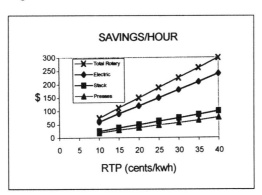

Substantive savings were achieved by '"working" the RTP rate by reducing consumption during high cost hours and increasing consumption during low cost hours. However, *major* savings resulted more recently, when a new electric melter was installed and electric consumption increased approximately 50%. This has resulted in consumption far above the CBL, and very desirable incremental costs for all the increased load, with little regard for timing.

THE FUTURE

At the time this text was originally presented, deregulation procedures in Pennsylvania were about to be finalized.

RTP rates have certainly been a harbinger of the future deregulated world, because of the metering and software which allowed hourly pricing data to be received by end users the day before implementation, and which allowed hourly load data and actual load profiles to be generated and used by RTP end users.

CHAPTER 14

GAS PURCHASING—
BUSINESS, LEGAL, AND
CONTRACTING ISSUES

"Competition"	"Market Pricing"
"Choice"	"Marketers"
"Restructuring"	"Aggregators"
"Deregulation"	"Unbundling"

These are the buzz words of a new era in the energy business. While the daily newspapers and radio are full of articles and advertisements concerning the impending opening up of the electric industry to customer choice of supplier, little attention is being given to choice in the natural gas markets.

The fact is, however, that except for California and some scattered pilot programs, retail electric markets are not yet open and even the imminent deadlines for retail access may prove to be illusory. For example, Rhode Island retail electric markets, by law, were open to competition July 1, 1997 but a month later less than a handful of customers had chosen alternative suppliers. In contrast, customers everywhere are already able to choose their own gas supplier, other than the local gas utility that had supplied all customers for so many years.

Presented in *Cogeneration and Competitive Power Journal*, Vol. 13, No. 2, by Eric J. Krathwohl, Esq.

With this new freedom of choice comes not only a number of benefits, but also risks. This chapter seeks to provide customers some guidance in obtaining such benefits and avoiding the risks. Ultimately, that is accomplished through a careful selection process, best done by means of an RFP with expert assistance, and through a negotiated gas contract. Before addressing specific contracting issues one must understand the legal and regulatory framework which governs the transportation of the natural gas.

LDC TRANSPORTATION ISSUES

Although a customer generally is able to choose a supplier other than its local utility, often referred to as a local distribution company, or LDC, that customer will still have to take transportation service from that LDC for delivery of the gas supply from the city gate to the customer's delivery point.

Such transportation service is governed by tariffs filed with the state utilities commission. Generally, such service is available to all commercial and industrial customers and, in some cases, residential customers. Some LDCs, however, have not updated their transportation tariffs and may restrict transportation service to customers that have a certain level of usage. Also, the terms and conditions applicable to LDC transportation can be very important to the feasibility of the customer shipping its own gas.

The primary area of risk to a transporting customer and the biggest area of dispute concerning LDC transportation terms and conditions are those provisions concerning balancing of gas supplies delivered and consumed. Such repeated or significant imbalances can lead to substantial penalties by the LDC. Such penalties can be straight charges, very high prices for overtakes (essentially a sale of gas by the transporting LDC, very low prices for undertakes (gas essentially sold back to LDC) and additional charges for costs that the LDC may incur that result from such imbalances.

Such charges are well accepted and must be considered in the choice of type of transportation service, the type of supply service and even the supplier itself. Some transportation customers even contract with the LDC for transportation quantities much greater (e.g., 20%) simply to avoid the risk of imbalance penalties.

The nature of transportation service and the balancing provisions is relevant to choices of services and suppliers as follows. Depending on the factors such as price for different types of transportation service (e.g. with or without built-in balancing service) and the availability and price of any separate balancing services, a customer may wish to enter a gas supply arrangement where the marketer agrees to take responsibility for all balancing requirements and has the operational capability and financial strength to do so.

Other characteristics of transportation service that may be important to a customer arranging for its own supply are those concerning pipeline capacity release (i.e. long line capacity for delivery of supply to the city gate) and availability of storage service. Both are relevant to the gas supply arrangements.

For example, a voluntary capacity release to former LDC sales customers that arrange their own supply allows significant flexibility to customers and interested marketers. Some marketers may want the capacity while others may not need it.

On the other hand, if an LDC has a mandatory capacity release program, that fact may restrict the number of interested marketers or the prices they can reasonably bid. In fact, in Massachusetts there is a pending proceeding on that very topic: how capacity release implementation has created hurdles to competition.

All natural gas market players must recognize the implications of local transportation terms and conditions, such as balancing and capacity release, and plan their transactions accordingly. Because of that importance of the LDC transportation tariffs, all parties should participate in regulatory proceedings (and even legislative efforts) that are relevant to the developing competitive markets.

One other service that an LDC may provide is storage service (or "virtual storage service"). Storage service can allow a customer greater flexibility in the amounts of gas and type of service for which he contracts. Regulatory agencies are increasingly looking at such services as a potential means of aiding competition and choice.

SPECIFIC GAS PURCHASE CONTRACT ISSUES

Pricing

Pricing, of course, is the central focus of any customer's purchasing decision. Determination regarding which supplier is offering the best price and which pricing option from a given supplier is most advantageous requires some degree of expertise and experience with such matters.

Considerable education may be obtained from the suppliers themselves, but whether a customer's purchasing manager can efficiently distinguish among different offerings and make the best supply choice is questionable. Especially, if that customer has not had prior experience in purchasing its gas supplies, the assistance of a consultant can be very helpful for customers that lack the expertise in their own staff.

One question that customers who are new to the energy purchasing process raise is the extent to which they are at risk for price fluctuations. If the customer has entered a fixed price contract, they should not have such risk. If the customer specifically bargained for exposure to price risk, he should benefit from a lower price, at least initially. One might legitimately fear that a supplier who has agreed to a price that becomes very favorable (to the customer) relative to the market may seek to renege on the contract. From the legal perspective, that should not be a concern, assuming a well-drafted contract.

Nor should the supplier have an economic incentive to act in that manner. Most suppliers that a customer ought to be doing business with would have hedged their price upon entering the contract. By hedging its price for a given supply at the time of contract entry, the marketer is insulated from the market risk the customer fears will trigger a breach. The nature of such hedging may be a legitimate point of inquiry for a customer. Contracting issues such as the time period for which a price is available (usually one year for a one year contract) and load level for which a given price applies may also exist.

Balancing Issues

As discussed above, the preferable approach from the purchaser's perspective is to have its seller undertake all responsibilities relative to balancing. In fact, this is more the norm. In those cases, the marketer will make all arrangements for transportation of the supply to the customer meter to eliminate all such efforts for the customers. Note also, that the marketer with a larger portfolio of customers will be in a much better position to balance the load than the customer.

Where the marketer undertakes balancing responsibilities, however, the marketer may reasonably seek: (i) detailed historical usage information; (ii) advance notification of any operational changes that would result in changed gas consumption, and other relevant information such as customer plant outages. Flexibility regarding these issues may be a major negotiating point. For example, a marketer will seek to impose costs on the customer that result from the customer's failure to provide adequate information.

Contract Volumes

The basic quantity of gas a customer contracts for is largely a business/operational issue. To the extent, however, that a customer foresees the possibility of significant changes in its load (e.g., due to growth, relocation etc.) it should negotiate for flexibility. Such flexibility can result in concrete savings to the customer.

Nature of Service

A customer may take firm or interruptible service, or any variation of either (e.g. 330 day firm). The customer's operational needs and capabilities will be the basis for this determination, but the different levels of service should bear different prices. Knowledge about the supplier's needs and resources may be helpful in negotiation.

Also, if service is firm, the contract must reflect that and the customer should have elicited information in the due diligence process to be confident that the supplier has adequate capabilities to fulfill that commitment to firm service.

Liability and Indemnification

Another contracting/legal issue that can be a fairly significant concern is liability and indemnification. Following the sound rule of allocating risk to the party best able to control that risk, the general rule, as in most gas contracts and tariffs, is to allocate liability for any damages to the party whose control the gas was in at the time of a problem.

Similarly, liability may follow title to the gas. Some contracts will provide for indemnification by one or both parties. Such matters can be negotiated, but in any event a customer should be aware of the extent of any indemnification required by the contract. Some contracts will be of even greater concern in that they continue the long standing utility policy of freedom from liability for damages. It is unlikely that a seller would agree to assume liability for consequential damages, but a customer should be aware that its supplier will assume only limited risks.

Billing and Payment Issues

Typically, marketers have strict provisions regarding time of payment and application of interest in the absence of timely payment. Customers should take care to ensure that such provisions which comply with applicable law are not overly onerous.

Miscellaneous Issues

Some contracts may require a customer to pay for metering equipment. The extent to which metering costs will be an issue depends partly on size and type of supply and nature of balancing services that a customer takes. Smaller customers, or customers that take a comprehensive balancing source may not necessitate complex metering and can avoid such costs.

Also, some competitive suppliers see metering and associated information systems as a means of offering more services, (e.g., load management) and thereby distinguishing themselves from competitors. A savvy customer may be able to negotiate some benefits in this area.

One contract provision that requires careful attention concerns provision of the customer's electric supply. Marketers frequently insert into gas contracts some obligation on the customer relative to a future electric supply contract. Such a provision may be as innocuous as the grant of a right to the gas supplier to participate in an RFP, or as anti-competitive (and potentially illegal) as requiring the customer to purchase electricity from the same seller once retail choice was available for electricity.

SUMMARY

These are but a few of the issues relevant to the gas purchasing efforts in the new era of customer choice. As the marketplace develops and as the sophistication of buyers and sellers increase, it is likely that many new forms of service and pricing will develop and the associated contracting issues will also increase.

CHAPTER 15

THE ROLE OF
THE GAS MARKETER

The role of the energy marketer has steadily evolved over the last decade, largely due to the deregulation of the natural gas industry. Beginning in the early 1980's, the catalyst of this new trend was the unbundling of Fortune 500 Industries. As Orders 200 through 636 unfolded, and the fast-paced onset of electric unbundling with Order 888 was realized, the role of the marketer has taken the shape of what now can be called your "Total Energy Provider."

From the beginning marketers are asked to wear a myriad of hats. Sales ability, and the experience of the individual to capture a market share in a given region, is their chief responsibility. The marketer must be knowledgeable of the tariffs that apply to different end users associated with their respective LDCs.

This information is utilized in order to implement and make the best recommendations based on specific customer needs.

In addition, keeping abreast of changes in the regulatory environment is also an important element of successful marketing. This allows a market to provide the customer with up-to-date information about how changes in regulatory affairs will affect their business. All these components combine to form a successful marketer who knows that service is key.

Service pertains to nominating the correct amount of supply, balancing accurately, understanding your customer's daily opera-

Presented in *Cogeneration and Competitive Power Journa*, Vol. 13, No. 2, by Deborah R. Daily.

tions, and knowing the type of equipment your customer uses. A marketer's service is only as good as his ability to properly communicate information on market price, supply and demand and weather projections, which all tie into a package that provides relevant and reliable information to the customer. It is service that ensures repeat business, as well as customer referrals.

In today's market the company that stands behind the marketer should offer services that the customer needs. Today's energy purchasers know more about the market than ever and it is imperative that marketers respond to their initial questions with intelligent responses. In choosing a "total energy provider" company several questions are asked by end users. What is the financial strength of your company? Does your company own supply and/or pipeline? Does your company have other hard assets such as electric power generation? If so, is it located within the customer's region? Does your company have local representation as well as a history within the region?

How strong are your risk management programs? What other services can you provide? Does your company have expertise and in-house staff with the ability to offer and finance projects, such as boiler conversion, cogen, conservation and efficiency, load tracking along with total energy management?

If a customer is only interested in price, then he must be made aware of what transpires behind the scenes of a typical contract signed between the buyer and seller. The support staff is responsible for creating a seamless transaction. This staff is key in making the transaction of moving supply and product from the "wellhead to the burnertip."

In order to produce one bill, several transactions are required within a company. The buyer supplies the most competitive price available on the market. A Transportation & Exchange (T&E) representative nominates that supply at the wellhead. Another T&E rep nominates a portion of that supply to the respective pipelines in order to deliver to a particular city gate and still another T&E rep handles the nominations and balancing of that supply behind the city gate.

At the end of the month gas accounting matches supply vs. actual pipeline delivery, the marketer checks for accurate pricing and volume and the risk management desk clears any hedge mechanisms. Accounts receivable creates an invoice and is responsible for collecting funds from invoices due. In turn, accounts payable is paying out to the suppliers and pipelines.

In addition, the Marketing Services Representative must be able to answer customer questions and concerns. In total, it takes at least ten staff members to complete each and every transaction and it is the marketer who is responsible to the customer for ensuring the accuracy of all these processes.

As we move forward into the wheeling of electricity the marketer must be prepared to offer a myriad of packages. Electricity is three times the size of the current gas market; this in turn will be a large component of a customer's energy portfolio. Therefore it is imperative that the customer is made aware of the many options now available in the market place.

For example at The Eastern Group we have created a risk management product that allows a customer to use his electricity yet pay a price based on the prevailing gas prices, oil price and/or electric price, whichever is cheaper within his region. Load management is a key component to electric wheeling. Whereas natural gas is a 24-hour exchange, electricity is every 15 minutes. The marketer must be fully educated in this area to best service his customers. Overall the marketer must have the tools in house to offer a total Btu package.

In order to communicate each of these transactions within a company, information must be made accessible to the marketer. The Eastern Group has the latest technology in a desktop management system. Our Contact Management System allows all intracompany communications to be posted behind a specific customer's account. In our Gas Management System, each customer file is updated daily. This provides the marketer with a valuable tool to monitor nominations and invoices, as well as receive messages left by accounts receivable. Once in receipt of this kind of information, the marketer is better equipped to follow up with products being implemented on a particular project, customer conversations with marketing services, follow-up calls and a long list of other variables which are inherent in today's marketing environment.

Currently marketers in the northeast regions must have a market niche in which they possess the ability to deliver supply. Due to the winter heat demand, pipeline capacity can be constrained if proper contracts are not put in place. This is a possibility whose implications every customer should understand.

The word "guarantee" is a misnomer in this industry; *forces majeure* have taken place in many unforeseen incidences. As pipeline capacity increases over the next few years many of these concerns will eventually dissipate. In regards to the electric industry currently many unknown factors remain that can leave a large margin for error.

The best recommendation I can offer my customers is to begin with a load management system following all fuels. This technology will track actual usage and allow us to make recommendations to our customers on how to increase efficiencies and conservation as well ways to lower demand cost.

The next time you look for that "nickel deal" be ready to get what you pay for. Customers who recognize the need for outsourcing their energy management must deal with a company that can offer a seamless approach to "Total Energy Solutions." This way the customer can position himself for the most attractive package in a deregulated market and be competitive within their industry.

CHAPTER 16

COGENERATION:
WHERE WILL IT FIT IN
THE DEREGULATED MARKET?

S everal states in the United States are opening their electric
power markets to competition. Among them California and
Massachusetts on Jan. 1, 1998 (currently California experiences a problem with the trading computer's communication, so
the implementation has been delayed beyond Jan. 1, 1998), Rhode
Island on July 1, 1998, Pennsylvania on Jan. 1, 1999, and Michigan
will phase in competition through 2002.

Cogeneration due to potentially high efficiency can be very
competitive in a deregulated market. Cogeneration can achieve
extremely high levels of thermal efficiency, much higher than the
most advanced and sophisticated combined cycle power plants
generating only electric power.

And thermal efficiency is one of the key factors in determining the power plant economics and feasibility. High efficiency
means a lesser amount of fuel is used to generate the same
amount of energy. In turn, burning a lesser amount of fuel means
that fewer pollutants will be emitted.

PURPA 210 (Public Utility Regulatory Policy Act, enacted in
1978) legal definition of cogeneration reads: "Cogeneration means
the sequential use of energy for the production of electrical and
useful thermal energy... subject to the following standards:

Presented in *Cogeneration and Competitive Power Journal*, Vol. 13, No. 2, by Moisey
O. Fridman, Ph.D., P.E.

a)   At least 5% of the cogeneration project's total annual output shall be in the form of useful thermal energy.

b)   Where useful thermal energy follows power production, the useful annual power output plus one-half the useful annual thermal energy output equals not less than 42.5% of any natural gas and oil energy input."

In 1981 an amendment was issued that reads: "Cogeneration technology means the use for the generation of electricity of exhaust steam, heat, or resultant energy from an industrial, commercial, or manufacturing plant or process, or the use of exhaust steam, waste steam, or heat from a thermal power plant for an industrial, commercial, or manufacturing plant or process."

The legal definition of cogeneration does not lend much understanding of its nature to a person who is not familiar with cogeneration principles. Besides, the legal definition is misleading and lacks clarity. The right definition should be: **cogeneration is the simultaneous generation of electric power and useful thermal energy from any source of fossil or nuclear fuel.**

The thermal efficiency of conventional power plants is low due to significant energy losses that cannot be avoided unless a sophisticated and more expensive cycle is designed. At a steam turbine plant the loss occurs during the steam condensation where the energy, equivalent to the latent heat, is rejected to the atmosphere via cooling towers. In a combustion turbine open cycle or a reciprocating engine application, again, a major energy loss occurs with the exhaust gas discharged to the atmosphere at high temperatures.

A typical utility steam turbine power plant has a 32% to 38% net efficiency (net efficiency is the ratio of net power generated to the fuel input. In turn, net power is the power at the generator terminals minus auxiliary load, sometimes called parasitic load). Table 16-1 shows the net efficiency of some steam turbine utility plants included in the top 25 U.S. plants for 1995. As can be seen, the best efficiency was within 36% to 38%. Combustion turbine open cycle efficiency is below 30%, and a reciprocating engine may have about 40% efficiency.

Let us now take a look at a conventional system for steam and electric power supply. Figure 16-1 shows a low-pressure boiler that supplies steam to process. This boiler is usually located at the same site where the user of steam is. The electric power is supplied from a plant (Figure 16-2) that can be located hundreds of miles away and the power is transferred via transmission lines.

In a combined cycle plant the gas turbine drives an electric generator. The exhaust gas is directed to a heat recovery boiler (a common term HRSG) and then exhausted through a stack. The HRSG generates high-pressure steam that is introduced to a steam turbine coupled with another electric generator. Sometimes, medium- or low-pressure steam from the HRSG can be inducted into the steam turbine to boost its power output.

The reader can see why a combined cycle has a higher effi-

Rank	Plant Operator	Heat Rate	Efficiency
	(Plant)	(Btu/kWh)	(%) ★
1	Tennessee Valley Authority (Bull Run)	8,975	38.0
12	Pacific Gas & Electric Co (Moss Landing)	9,441	36.1
25	So. Carolina Generating Co. (Williams Station)	9,580	35.6

★ Efficiency = $\dfrac{3413 \text{ Btu/kWh}}{\text{Heat Rate (HR)}}$ x 100 %

Table 16-1. From the Top 25 Heat Rates (HR) at US Steam-Electric Plants List—1995)

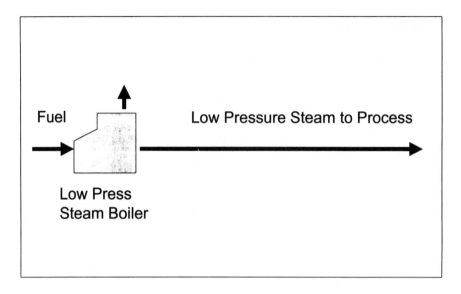

Figure 16-1. Conventional Process Steam System.

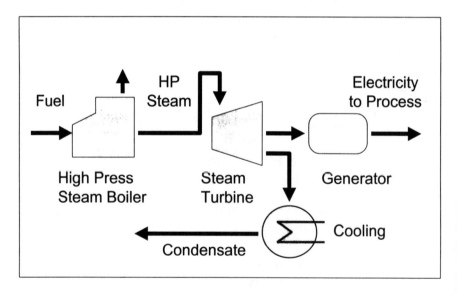

Figure 16-2. Conventional Electric Generating System.

ciency than an open cycle gas turbine or a steam turbine. The energy of the exhaust gas from the gas turbine is partially utilized in the HRSG, and the steam turbine exhausts less steam to the condenser. A combined cycle may achieve about 55% to 60% efficiency.

Usually, in a combined cycle the gas turbine generates approximately 2/3 and the steam turbine generates 1/3 of the total power. If a combined cycle is used in a cogeneration application, e.g. when some steam from the HRSG or from the steam turbine extraction (refer to Figure 16-3), or both is used for process or heating/cooling, the ratio of power generated by the gas turbine and steam turbine can be quite different.

Of course, the efficiency of a combined cycle cogeneration plant with the same initial parameters will be higher than 55% or 60%. In general, a good cogeneration plant can achieve an 85% efficiency and above.

All the cogeneration cycles with backpressure (non-condensing) steam turbines, condensing turbines with extractions, com-

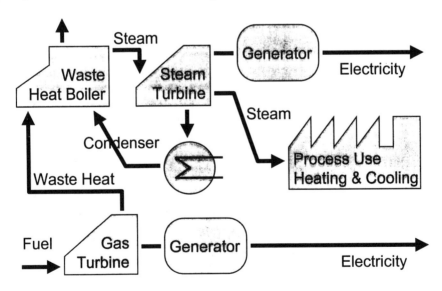

Figure 16-3. Cogeneration—Gas Turbine Combined Cycle

bined cycles or heat utilized in a HRSG from a combustion turbine or from a reciprocating engine are called *topping cycles*. A topping cycle is the most common arrangement for cogeneration. When the fuel is first used for process and then the thermal energy of the exhaust gas is utilized for power generation, such cycle is called a bottoming cycle.

Another emerging technology for cogeneration application may be fuel cells. Their efficiency ranges from 40% to 57% depending on the type of fuel cell technology (phosphoric acid, molten carbonate or solid oxide). Again, in a cogeneration arrangement such efficiency can rise significantly.

Despite the variety of arrangements, **all the topping cycle cogeneration plants have certainly one thing in common: they supply useful thermal energy.** And when we show the steam extracted from a turbine or from a HRSG we have in mind either direct application of steam, or its use in a heat exchanger to produce hot water, or in an absorption chiller to produce chilled water. The aforesaid brings us closer to understanding the nature of the cogeneration plant's economics. And good economics are the key to survival: if a plant is highly economical it is competitive.

IMPORTANCE OF THERMAL LOAD AVAILABILITY

Depending on the type and size of the thermal load and the type and size of the electric demand, various types of prime movers and various sizes and turbine initial parameters can be selected. Each option is associated with a certain capital cost and has corresponding operating expenses. By comparing all annual costs versus revenues generated, the best option can be selected. Such a process is called cycle optimization.

Figure 16-4 shows the relationship between a steam turbine specific power (kW-hr generated per million Btu of thermal energy extracted) and initial steam parameters and extraction pressure. For example, if you use utility compatible 2400 psig/1000°F as inlet parameters, and the extraction pressure is 100 psig, you can generate approximately 95 kW-hr/MMBtu. However, if a cogeneration plant has inlet parameters 600 psig/750°F, it will generate only about 50 kW-hr/MMBtu. Of course, it will cost less to build the second plant, but it will generate lesser revenue.

Let's now take a look at the legal definitions of cogeneration and at the 5% and 42.5% minimum requirements referred to at the beginning of this article. The 5% requirement is called operating standard, the 42.5% is called the efficiency standard. Using numbers from Figure 4 we can find that for a 600 psig/750°F plant and a backpressure turbine with 100 psig extraction pressure, the ratio of *thermal energy over the sum of electrical plus thermal energy* would be 85% instead of the 5% minimum required.

The 5% requirement means that approximately 10% only of the steam introduced to the turbine will be extracted, and 90% will go to the condenser with all the low efficiency implications. Which brings this cogeneration plant close to a conventional electric power plant, except, due to the cogeneration plant's low initial

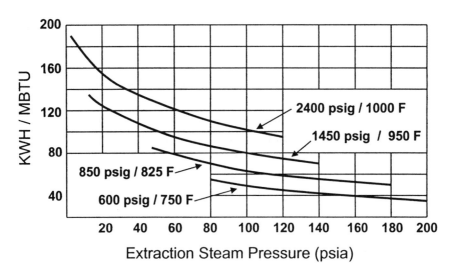

Figure 16-4. Steam Turbine Specific Power Output vs. Extraction Pressure.

parameters its efficiency will be even lower than that of an average utility plant.

Of course, having a large condenser will add flexibility to the operation: it facilitates load switching and some peak load shaving. However, such performance should be limited only to a few hours a day, otherwise the cogeneration plant efficiency measured on an annual basis will drop substantially.

Therefore, to limit the condenser operation, a second requirement, the "efficiency standard," was formulated which requires that on an annual basis the ratio of the sum of electrical plus one-half of thermal energy over the fuel input shall be equal or more than 42.5%. In other words, for short periods of time you can use your cogeneration plant just for electricity generation, however, on an annual basis you need to generate more useful thermal energy.

Again, this requirement still leaves opportunities for extensive condensing operations. In our example the ratio will be higher than 50%. Should we use different initial parameters, this ratio may approach 70%. In other words, a lower annual ratio means you have generated much more electric power than is stipulated by your thermal load, and, probably sold the extra electricity to the utility.

Some states did develop more stringent QF (qualifying facility) requirements than FERC (Federal Energy Regulatory Commission), for example, in Connecticut a 20% minimum operating efficiency was required instead of 5%.

During the 80's many developers that met the above referenced minimum QF requirements, enjoyed the tax benefits and gaseous fuel use benefits (including special cogeneration low tariffs) and built independent power plants. Unfortunately, it was found later that some of these plants did not have one important key factor that makes cogeneration feasible and beneficial—**the proper thermal load profile.**

Figure 16-5 shows a hypothetical daily thermal load profile (with emphasis on the thermal load, but if the plant is put in a dispatch mode, the electric load profile may significantly affect the plant economics as well).

For simplicity let's assume that this is the plant's typical process load profile which means this plant operates round the clock and that approximately half of the time the load is close to 50% of the maximum. If you take a typical daily cooling load profile, its shape will look similar to Figure 16-5, except the peak and semi-peak area will be much narrower.

It would be a mistake to size the power equipment (high pressure boiler and turbine) for the maximum hourly load, because the larger size equipment will cost much more, and will be operating at full load only for a few hours a day. It is much more economical to size the equipment for the base load only, and shave the peak load with a low pressure boiler for process steam, or with a TES (Thermal Energy System) in a cooling application.

The next simplified graph (Figure 16-6) shows how much thermal energy can be used by one-shift, two-shift, and three-shift operations that have the same hourly maximum load. It is clear that a three-shift round the clock operation facility will require much more thermal energy annually as compared to a one-shift, although their hourly maximum load is the same.

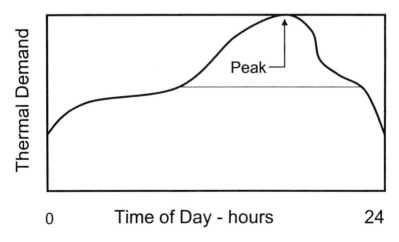

Figure 16-5. Process Load Profile—24 hr

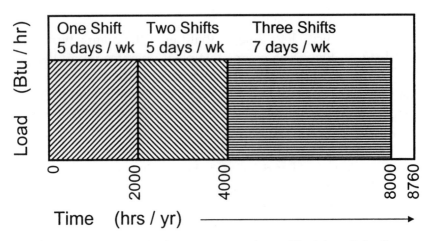

Figure 16-6. Annual Process Load Profile (simplified)

Therefore, it is much easier to justify a cogeneration plant that will supply thermal energy to a three-shift operation. If a facility operates only 1 shift, 5 days a week, in most cases the cogeneration plant will not be economical.

The next graph (Figure 16-7) shows an annual heating and domestic water load profile typical for a Midwestern city. The cooling load annual profile for a southern area (for example South California) may look alike, except there is no flat segment, and the word "winter" is replaced with "summer."

Again, the power equipment shall be sized for the base load only. The ratio of the optimal thermal load, for which the power equipment is sized, to the hourly maximum thermal load is termed "cogeneration design coefficient" (CDC). It is a very important factor for equipment selection and determination of the feasibility of a cogeneration plant.

As can be seen from Figure 16-7B if we select a CDC of approximately 0.5 for a heating or cooling load, this will assure that the cogeneration plant will be able to use over 80% of the annual thermal energy demand for simultaneous generation of electric power. The remaining peak load should be shaved off by less expensive low-pressure boilers or cooling sources.

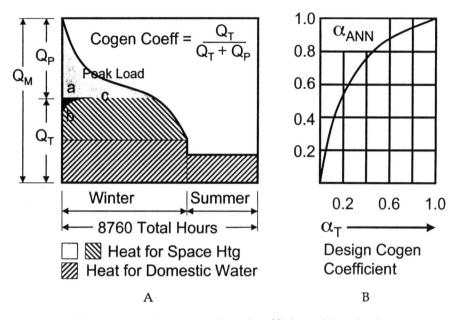

Figure 16-7. Cogeneration Coefficient (Heating)

Hopefully, the above examples illustrate the importance of understanding the thermal load profile the cogeneration plant developer and/or a potential plant owner should consider. If you do not have valid actual thermal loads (not "maybe, in the future") you are wasting your money for cogeneration. But if the loads are "in store" on an annual basis, your project can be a winner.

The author had experience with a cogeneration plant built in California that was designed to meet cooling loads of a famous hotel using absorption chillers. However, after the cogeneration plant was built, the owners of the hotel decided that they did not want absorption chillers due to their poor past experience with these types of chillers. Therefore, the cogeneration plant currently does not have thermal energy demand, and is mothballed. If the hotel owners were familiar with the cogeneration principles, this plant would either never have been built or would have been equipped with modern reliable absorption chillers that operate successfully.

COGENERATION AND DISTRIBUTED GENERATION

All of us have heard about distributed generation (DG) that is supposed to hit the market due to the deregulation. During recent years there have been a host of articles about DG; further elaboration isn't necessary. In the majority of cases the ideal DG plant size will range from small to 25 MW—50 MW.

An often-asked question is what size cogeneration is feasible. In this article only a general answer can be given, because the economics of a cogeneration plant are very site specific. The majority of the readers probably know how an effect of scale works. The larger the plant is, the lesser is the capital cost per kW installed within the same technology, say steam turbine or gas turbine plant.

At smaller plant sizes the technology may need to change, for example, the application of reciprocating engines or fuel cells instead of steam or gas turbines, or application of a recently emerging new breed of very small gas turbines in the microcogen range.

One problem for microcogen gas turbines, though, may be the fuel supply. The microcogen gas turbines may require a high compression ratio, which means a need of natural gas pressure in the range of 350 to 500 psi. This may add a significant cost to the cogeneration plant and in some instances make it hard to justify.

By the way, the term "microcogen" needs to be defined. According to many authors and publications, microcogen units are below 500 kW. However, California Assembly Bill AB 1890 page 90, paragraph 331f labels "microcogen" everything below 1 MW.

In general, the size of reciprocating engines can range from 20 kW to 10 MW and higher, fuel cells could range from 200 kW to 2 MW, micro gas turbines range from 50 kW to 500 kW, small gas and steam turbines are in the 1 MW to 5 MW range. Common medium size turbines and combined cycle cogeneration plants are in the 10 MW to 50 MW. Large units can range in several hundreds of MW.

Recent examples of moderate size cogeneration plants that

were found feasible: San Diego, California, 2.4 MW hospital co-generation upgrade to increase the capacity to 4.7 MW and add absorption chillers; Mojave Soledad Mountain Mine (California) 10 MW cogeneration development; Chicago, Illinois, McCormick Place 4.5 million sq. ft. three-building central heating/cooling co-generation plant with three 1.17 MW gas turbines.

As can be seen, the cogeneration plant is an ideal source for distributed generation. Cogeneration and the DG go hand-by-hand. So, being highly efficient and ideal for DG, will the cogeneration technology be competitive in a deregulated market? The answer is yes, of course.

OTHER ISSUES AFFECTING COGENERATION

Where stable thermal loads are available, proper selection of the major plant equipment will assure high efficiency. In turn, high efficiency of a cogeneration plant means lesser fuel used for generating the same amount of thermal and electric energy than can be obtained from a conventional scheme (thermal energy from a separate boiler and electric power from a utility plant).

And burning a lesser amount of fuel means lesser amounts of pollutants being emitted to the environment. In other words, a cogeneration arrangement is beneficial to the environment, and contributes to reduction in global warming.

The author feels that EPA should label cogeneration plants as "green" or "semi-green" technology and come up with an emission credit system that will serve as a bonus for generating a certain minimum of energy using lesser amount of fuel.

In December 1997 the United States Vice President Gore went to Kyoto (Japan) for an environmental summit on global warming. The United States was pressured to tighten emission limits further. This could have a devastating impact on energy prices and promote further the plants with the highest efficiency, among them, of course, cogeneration plants.

One extremely important question that may be asked by owners of existing cogeneration plants or by facilities that can be potential host sites for cogeneration: what if we do not have qualified manpower resources to operate our own cogeneration plant? There are several solutions. Companies like Steward & Stevenson, Kraft etc. and many local operators can sign a long-term operating agreement with such facility.

Another example of companies that are ready to take advantage of deregulation, is a recent agreement between DuPont Co.'s Conoco oil unit and a utility company (American Electric Power Co.) to set up a joint venture aimed at meeting the needs of heavy industrial users. This new company will buy the power plants of various manufacturing companies (steel, refined-oil, petrochemical and paper producers), upgrade them and then lease them back for a 17-year period. If asked, this company will also operate these plants.

One more example of this trend is the recent Tractebel's Power Inc. acquisition of two cogeneration plants from Simpson Paper—a 40 MW plant in Pomona, and a 42 MW plant in Ripon, both in California. These plants will continue to supply steam to Simpson Paper.

We keep hearing horror stories about the CTC (competitive transition charge) penalties imposed by some states to recover the utilities' stranded costs and to facilitate the deregulation. However, these charges are temporary. For example, in California the CTC charge will end by June 30, 2000. For a well-selected cogeneration plant a 2-year surcharge of about 2 to 3 c/kW-hr is not fatal. Such plant can generate revenues that will easily offset the CTC losses.

The bottom line is: if proper thermal loads are in place, and if the cogeneration unit is configured and sized to fit the available loads, and the proper technology is selected that will best meet the electric-to-thermal load ratio, such cogeneration plant will be a viable and competitive plant in a deregulated market. Such plant will save a significant amount of fuel and ultimately contribute to reduction of global emissions.

At the beginning of this chapter, pros and cons of using a condensing steam turbine arrangement were mentioned. This feature may be an advantage for a cogeneration plant in a deregulated market, adding operating flexibility, that allows load switching. Cogeneration plants that are equipped with condensing steam turbines or combination gas and steam turbines may have load-switching capabilities.

During the Electric Peak hours it may become beneficial to generate more thermal power from low-pressure boilers, TES, etc. and to divert more steam to the condenser instead of the extraction, in order to increase the electric output and sell extra peak power at an instant high rate.

Such thermal-to-electric load switching can be partially achieved within the existing PURPA limits or, if PURPA is repealed, as some experts suggest, under modified rules. Of course, a condensing arrangement requires higher initial capital cost and some additional operating expenses, however, deregulated market economics may justify less efficient condensing mode part time operation.

Another important issue is the type of fuel to be used for cogeneration plants. There are large industrial cogeneration plants that are burning coal, and there are medium size cogeneration plants burning wood or biomass, or other exotic fuels. However, the majority of cogeneration plants are using natural gas.

Natural gas is a clean fuel, it requires lesser pollution control equipment. And since gas market deregulation there are no shortages in gas supply, and the prices keep falling. Natural gas pipelines are usually available in the vicinity of dense thermal loads, although modifications and/or pipe size increase may sometimes be required.

How do states view cogeneration? It partially depends on the state population density, and prevailing industries and utilities' attitudes. Many states encourage cogeneration. For example, the state of California AB 1890, Section 372a states the following:

"It is the policy of the state to encourage and support the development of cogeneration as an efficient, environmentally beneficial, competitive energy resource that will enhance the reliability of local generation supply, and promote local business growth."

So, where does cogeneration fit in the deregulated market? It's like our local store. We do not need to drive on a freeway or take a bus, shopping for lower prices. It is near us, and in many cases may be the most economical source of both thermal and electric power.

CHAPTER 17

ELECTRIC & GAS CUSTOMER CHOICE OVER THE WEB

A s electric and natural gas utilities around the country go through the laborious process of unbundling their rates and services and allowing their customers the option of selecting alternative energy suppliers while maintaining transportation services, many traditional daily operational aspects performed routinely by each utility must now be made available and/or shared with a host of old and new energy market players. These daily operations must ultimately be fully automated and consistently applied across a broad spectrum of new and aging utility information systems, or the administrative burden and transaction costs of unbundling alone will prevent complete, down to the residential customer, open market supplier choice competition.

But by simply improving the efficiency of the daily operations and transactional information flow between energy supply marketers and the utility using the Web, and reducing transactions costs, much more is required to make this a truly competitive and vibrant energy marketplace. Highly contentious issues related to daily load balancing, customer load aggregation rules, peak day reliability, consumption metering, nomination/scheduling rules, and many more, still need to be resolved. Even when all of the above are settled, if customers don't see any real savings from unbundling—because the "stranded costs" of uneconomic generation

Presented at the Competitive Power Congress '98 by David E. Molzan.

and gas supply pipeline assets are eventually all passed back to the customer for instance—it could take years before retail supply competition benefits reach all natural gas and electric consumers.

However, the Web will continue to add value in solving or mitigating these business problems specific to the unbundling of the natural gas and electric industry. It will also provide leading edge companies the tools to prosper and deliver even more value to their customers, in realtime[1] and accelerate the transition to a competitive energy marketplace.

WEB SOLUTION—CUSTOMER CHOICE APPLICATION

Tech 2000 Worldwide, Inc., has developed Web-enabled applications as an efficient means for a host gas[2] or electric utility to administer their retail customer choice programs. Energy marketers are able to enroll/switch/remove customers in customer choice programs, receive pipeline/supplemental nomination or power scheduling information and even exchange billing information over the utility's Web site. The Web interface for secure transactions between dozens of energy marketers and the host utility has become the electronic medium of choice for innovative "wires and pipes" utilities which see the ultimate goal of providing market-driven supply competition the best value for their customers and a sound business strategy for their stockholders and employees.

The Web allows an easy-to-learn and easy-to-use interface. All the energy marketer needs is an Internet service provider to access the Web and a Netscape or Internet Explorer Web browser installed on their PC.

Enrollment/Energy Supply Marketer Elections

One of the primary functions for the utility to handle on a daily basis is the administration of customers' migration from a utility's bundled sales service to its unbundled transportation service. This naturally requires that the host utility and each participating energy marketer know exactly which customer belongs to whom. Literally hundreds of energy marketers may be offering existing bundled sales customers upstream electricity or natural gas supply services.

A perfect customer match must be verified and approved between each energy marketer's contract list and the host utility, in order to begin many of the key operational functions. Based on many cases in the natural gas industry, which has been slowly unbundling for several years, this normally requires several faxes or phone conversions or e-mail, a tremendous burden to the utility and each marketer. Errors often occur in the exchange of this information. New customer file management systems or simple spreadsheet models have been hastily built to augment aging legacy customer information systems (CIS) that have further complicated the problem.

As is the case with many pilots around the country, most restrict the amount of customers, or the level of capacity available to competitive suppliers. The administrative burden will continue to grow with added complexity, confusion and delay, as more customers (from large commercial/industrials down to the smallest residential energy consumer) are allowed to migrate from bundled sales to transportation services from the local utility.

If you throw in the option of allowing these customers to switch suppliers once a month or once a year, and additionally allow them to come back to the local utility's standard offer bundled sales service—in a sense rejecting the unregulated commodity services from an energy marketer—you have a major administrative nightmare on your hands.

Account and Meter Number Transfer for Enrollment

The key variables for information exchange to accomplish unbundling program enrollment are the account and meter numbers that uniquely identify each utility customer. This information ties in the address, phone number, rate code, billing information, consumption history, load profiling information for monthly billing estimation routines, and the dozens of other fields unique to that account record.

The host utility Web site can efficiently and effectively manage this information transfer for the confirmation process better than with paper signature enrollment forms. Many pilots allow only paper confirmation of supply offer agreements from the energy marketer with a customer's signature. Besides the lack of quick turnaround, errors by the marketer or the customer in completing the form, or the utility staff in entering data, cause additional delays in verification and enrollment of customers. The customer is also asked to do more than is really necessary.

Paperwork transactions for choice enrollment will not be the future for the unbundling of the electric and natural gas industry. Just as anyone can order products and services over the phone or Internet with a credit card, so should the energy-consuming public. The fear of slamming (marketers switching customers without their permission), as in the case of early phone deregulation, will not occur for several reasons. The utility will be in the position to easily track the offenders, up to the minute the infraction occurred, and stiff penalties will be applied. In addition, marketers who engage in this practice will undoubtedly be made known to the public and their competitors, surely killing a positive brand name recognition. A marketer self-policing mechanism would provide even less of an incentive for slamming. The state's energy commissions have been very active and extremely clear about consumer protection and the amount of contract disclosure information necessary for marketers to engage in sales activities to prevent any wide-scale problems. Probably just as important is that energy marketers must purchase upstream generation or gas supply capacity assets, up to a year in advance, to meet the requirements of their customers' loads. There is no incentive to sign up customers for supply services under false pretenses and eventually have to back out of these upstream contracts, with penalties, because customers were mislead or slammed.

Internet Sign-ups

Customers should be allowed to accept offers over the phone or the Internet. The energy marketer can easily obtain all the key variables from the customer for verification and enrollment, if the utility does not supply this information directly to energy marketers. The marketer's ability to obtain this information from the customer (via phone, post card, etc.) acknowledges acceptance of their supply offer and allows the marketer to act as the customer's agent. It should be up to each marketer what paper transactions they desire between themselves and their customers.

Knowledgeable customers are the best customers, as we have all heard and believe. Communication about supplier choice that identifies key consumer benefits should also explain what actions they need to take in order to select an energy marketer. Marketers will behave in their own best interest, but more importantly will want to meet or exceed their customers' expectations with superior service. Energy marketers should be able to act as their customers' agents and take complete care of arranging for service with the utility.

The marketer, acting as the customer's agent, begins the enrollment process by simply entering an account and meter number for each customer in two required fields on a Web enrollment form. The data entry of one customer, or hundreds throughout the day, creates a separate "pending" enrollment file for each marketer that is sent off to the utility for processing against the CIS database. A matching routine is performed on a nightly basis, but could be done more frequently if necessary. These "pending" enrollment files will either be "confirmed," placed in a confirmed file or "rejected," placed in a rejected file. The confirmed customers are assigned a code by the utility to identify them as transportation customers. This code is also used to identify which marketer they belong to. If the account or meter number is incorrect or the utility has other criteria for rejection, such as non-eligibility for enrollment due to nonpayment, or wrong service territory, or simply the account number does not match any current customers in the database, the customer is rejected. Both the confirmed and re-jected customer files are sent back to the site under secure access for each marketer. Marketers can then download these lists into a database or spreadsheet to maintain their customer files.

Switching Customers to Another Marketer

The utility also has control over how, and how often it will allow a customer to switch suppliers. Because the utility has a marketer code to identify a customer for transportation service, it can choose to allow that code to be replaced. So if a marketer at-tempts to enroll another marketer's customer, the host utility can control the rules of the game. A marketer can go after current bundled sales customers, or if they desire, current transportation customers. A marketer can enter the same account and meter in-formation for an existing customer, and the utility can simply re-place the old marketer code with the new marketer code. Addi-tional steps have to be taken to notify the losing marketer they no longer have this customer. The creation of a "Switched" list on the Web accomplishes this and helps the marketers control their up-stream assets, as well as marketing efforts. The message will show they have lost customer x to another marketer. The utility can also follow up with a programmed letter to the switching customer stating the change. The utility will offer a toll-free telephone num-ber to call in case of error. Again, there is no benefit for slamming because of the known and measurable entry times for monitoring this activity.

Switching Customers Back to the Utility

The switch may also be made back to the utility. If a customer wants out of the program they may contact their supplier (or their utility) over the phone, based on the customers' contract provi-sions with their suppliers. The marketer may have other reasons to remove a customer, such as to satisfy the customer. Again, the marketer acts as the customer's agent and simply "clicks" on a "remove" button, rather than an "add" button, after entry of the account and meter number. The utility replaces the marketer code with a blank or a code that identifies the utility as the supplier (thus the customer returns to bundled sales service).

Action by the utility as agent for the customer, will also show up under the switched button, as customer x has been switched back to the utility. The marketer has the ability to track confirmed, rejected and switched lists, updates on a daily basis.

Web Customer Record Exchange Benefits

This customer-records information exchange, between the utility and all participating marketers, flows through one central and focused location. Information is processed automatically, rather than each marketer mailing, faxing or emailing lists on a sporadic basis to different utility personnel, with different file formats, at different times.

The marketers are also able to track their market share on a daily basis when the total number of customers enrolled is posted. If a limit on the number of customers, a specific rate class, or a level of capacity is close to being reached, the commodity marketers can stop their marketing efforts so as not to spend any more money to solicit customers that they won't be able to enroll anyway. This automated realtime posting of enrollment information is a simple and effective tool, much better than waiting a couple of weeks to see where they stand on utility imposed competition limits.

Billing Information Exchange—
One- vs. Two-Bill Options

The exchange of this key information between the energy marketer and the utility also facilitates management of different billing options for the customer. One bill—combining commodity and transportation, or two bills—one from the commodity provider and one from the utility.

The utility may still desire to retain control over the transportation portion of the bill, so in some cases the customer is receiving two bills. However, two bills will not be the future for this service. Ultimately the customer will receive one bill from the energy marketer that will combine the commodity and transportation portion of the bill. The marketer will probably have systems that will send a single bill to a customer who has facilities all over the country. This additional value-added service is exactly one of the future benefits of unbundling services. The transfer of rates and meter readings between each marketer and the utility is also provided rather easily over the Web.

Utility Retains Billing Function

If the utility continues to do the billing—as in the one-bill option—the marketer simply adds a rate code, independently confirmed with each marketer, to the Web enrollment form and this information is added to the customer record as the commodity billing rate. The marketer will still need to receive the meter reading and consumption records for planning, so every night the utility sends back to each marketer a complete cycle billing record for those customer accounts that fall on that billing cycle.

Marketers Retain Billing Function

If the marketer does the billing for a one bill option, the meter readings, consumption data and all transportation rate information is passed onto the marketer. Energy marketers may not have a billing system in place to do the billing or may not want the burden of building or purchasing a new billing system for only a few thousand energy customers, probably until they reach a certain cost to benefit threshold. In the Bay State Gas Company choice pilot, the utility actually performed the billing services for several marketers and placed the billing information on the marketers' own letterhead. Bay State Gas wanted as many marketers to participate in the program and felt this one obligation, priced out at cost, would attract more marketers to the customer choice program.

Upstream Pipeline Capacity and
Supply Portfolio Management

For the Bay State Gas customer choice program, marketers were given an optional choice of selecting upstream pipeline capacity from the utility, based on a proportional share their customers would take as a sales customer. The cost for the capacity was at the full FERC regulated price. They could also choose to purchase pipeline capacity on their own. Bay State also offered a supplemental supply service for requirements above pipeline capacity, again offered on an optional basis. Due to the lack of pipeline capacity in the Northeast, the majority of the marketers chose some level of additional supplemental supplies from Bay State.

Bay State fixes the allocable upstream pipeline capacity share allowed once each month based on the design day of aggregated customers' load as a pool requirement for each marketer. So as customers move from sales to transportation, from marketer to marketer, or transportation back to sales, the utility communicates this information back to the marketers each day.

Utility Planning Improvements

Short-term supply planning for the utility's remaining bundled sales customers is better managed with enrollment information available on a real-time basis. Upstream capacity needs can be sold or transferred directly to each marketer that enrolls a customer, based on the customer load requirements, at specified planning intervals, for a mid-month auction, or even on a daily basis, if large enough to be operationally practicable. Capacity still held by the utility can be sold in the capacity marketplace, or optimized to meet current bundled sales customer requirements. This is extremely important with choice programs that enroll customers throughout the year, rather than during a fixed enrollment period. This capacity asset transfer will certainly minimize exposure to costs born by non-migrating customers.

Stranded Costs

Many unbundling programs around the country are wrestling with the contentious issue of mandatory vs. optional upstream capacity assignment during this transition to an open market. This issue has major impacts on who pays for unwanted or uneconomic assets already in a utility's capacity portfolio, and what portion can be collected as stranded costs. Some states have settled on billing mechanisms that allow utilities to collect a negotiated amount for stranded costs from existing utility customers. In the case of electric unbundling programs in California and Massachusetts, however, these stranded cost payments have created little or no room for competing suppliers to offer lower prices to electric consumers. Because of this, very little action is taking place.[3]

Hourly Power Scheduling and
Daily Pipeline Nominations/OFO orders

Both gas and electric customers have historical monthly consumption information in their customer records, over a period of twelve to twenty four months, that provide information for forecasting future energy consumption. Electric consumers generally fall within a certain rate class that assigns them a specific hourly load profile—great information for short term planning purposes. Natural gas consumers usually have base load and heating degree day coefficient based algorithms that enable estimated billing procedures. In both cases monthly consumption readings are the standard form for billing customers.

The natural gas industry has been in the process of gradual unbundling for several years now, mostly with the migration of large commercial and industrial (C&I) customers to transportation service. The marketers for these large C&I customers must nominate, at a minimum, 24 hours in advance of their expected consumption to the Local Distribution Company (LDC) and also to the upstream pipeline transportation and gas supply contract providers. These pipeline nominations must be in the hands of the appropriate LDC supply personnel in time to manage the entire downstream distribution system for the next day.

Automated Meter Reading (AMR)

The transportation requirements for these large C&I customers usually requires daily telemetered consumption data to measure actual consumption compared to the nomination forecast. Large swings in daily over- and under-nomination forecasts, compared to actual consumption, are monitored to force marketers to meet their customers' load requirements as best they can. Penalties by the LDC are enforced to protect the bundled sales customers from subsidizing the transportation customers, especially during peak day or critical times, on a customer by customer basis.

As the number of customers migrating to transportation services are increasing it has become necessary to "pool" loads by marketer and manage daily balancing through the pools of customers rather than on an individual basis. LCDs are also allowing trading of imbalances between marketers. The reason for trading of imbalances is based on the fact that if the LDC's system integrity, in total, is not being "harmed," there should be "no foul" or penalty paid.

The cost of installing remote AMR equipment is also becoming a roadblock to further unbundling efforts. The smaller customer loads can and should be aggregated to larger pools for daily nominations. The vast majority of customer loads have been daily/hourly forecast by LDCs for years, based on aggregated loads for operational planning. Algorithms and load profiling information on an individual customer basis can be predetermined and managed effectively even with large-scale migration of customers to unbundled services.

Several unbundling programs, including Puget Sound Energy and Bay State Gas Company, provide the energy forecasts, in a sense, directly to each marketer, based on proven planning tools of the LDC. Many utilities, such as Pacific Gas & Electric Company, are further refining these forecasting models to corroborate these methods with residential AMR devices.[4] This will provide a much more cost effective means to mitigate daily/hourly systems integration and streamline the nomination/scheduling communications between the marketers and the utility.

Puget Sound—Hourly Power Scheduling

The Puget Sound site—www.psechoice.com—provides the enrollment and billing applications mentioned above, but also provides hourly power scheduling information securely to each participating energy marketer. These are 24 hourly megawatt forecasts provided each day on a one- to four-day forward notice, representing the most current pooling of customers for each marketer. Hourly load profiles and rate schedules are known for each customer as part of their customer billing record. The dynamic posting of hourly power schedules provides the energy marketers with one less operational issue to confront, and resolves some of the problems concerning timing of the postings and coordination with the host distribution utility.

Bay State Gas—
Daily Pipeline and Supplemental Nominations

The Bay State Gas site for unbundling operations—www.bgc.com—provides daily nominations, one to four days out. The amount of upstream capacity selected by the marketer, either from the utility or the marketplace, also determines the level of supplemental supplies that are nominated and provided by Bay State. The individual customer base load and heating coefficient algorithms are summed by marketer, by pool, by separate gate station delivery location and multiplied against the daily weather forecast feed. This information is updated every morning by 10 a.m. Operation Flow Order (OFO) information is also provided, should the system integrity be in a critical condition, as in extreme cold days in the Northeast. The nominated volumes most clearly represent the best load forecast for the pool of customers, and are updated based on the size and mix of customers in each pool.

Essentially, all daily nomination and power scheduling could be handled automatically over the Web with models and or load profiling tools. The results would be linked to upstream transportation management tools and downstream distribution planning systems to automate daily transportation monitoring and billing.

The marketplace will then drive additional value added services, such as with AMR, providing daily or even 15-minute interval consumption data for those customers who actually want to pay for these services. If the LDC had 1,000,000 customers to manage before on a daily basis, it can just as effectively manage them if 10 marketers had 100,000 customers each. Sharing their load profiling and/or daily forecasting models with each supply service provider will increase cooperation and coordination with the utilities "new" customer, the energy marketer.

EXIT OF UTILITIES FROM THE MERCHANT FUNCTION

In the natural gas business, many utilities are exiting the merchant function because they are not allowed to make a profit on the sale of gas. There can only be downside risk in maintaining that function in the gas business, as there are very few public utility commissions that reward for superior performance, and hand out penalties for poor performance. In the electric business, several utilities have begun exiting the merchant function and have sold off their non-nuclear generation assets.

These companies who are aggressively becoming "wires and pipes" regulated entities will most likely promote supplier choice initiatives as a means to improve customer services, increase market share and grow their regulated local transmission and distribution systems. Utilities that want to remain in the merchant function business will not aggressively promote supplier choice because of the supply difficulties in allowing customers to migrate between transportation and sales options.

If the consumer compares offers from various competitive

suppliers and also compares them with the local utility's "Standard Offer" bundled sales offer (still in the merchant business), there is usually little or no incentive for switching, because savings are minimal, at best, for the next few years.

CHOICE FUTURE

Customers value and expect choices in almost every product and service they buy, and will seek out the greatest value.[5] Most everyone expects that when an open market develops to choose an alternate supplier, this automatically guarantees lower prices for every energy consumer in the country. The reality is that, in some cases, a few residential customers may actually see prices rise with competition.[6] Some states have very low cost electric generation and gas supply assets. If these are sold on the open market, some out of state markets may outbid the local markets and sell them to higher cost markets for a profit.

Currently, 50% to 80% of a customer's energy bill pays for delivery of that energy (still regulated), depending on what part of the country you live. So the opportunity for commodity cost savings from an alternate energy supplier lowering your total overall bill by more than 10%, with less than 50% of the bill to play with, is probably not going to happen to many customers, any time soon. The California electric market is a clear example of what can happen when there is no commodity margin for profit. When you add to the marketers' equation the cost of soliciting customers, it is crystal-clear why this particular retail marketplace is a nonstarter.

Clearly this means small commodity savings alone will not drive customers to shop for cheaper energy supplies. Only huge electric or gas load customers, that can obtain a small percentage in energy savings, that amount to thousands of dollars, will participate.[7] Many energy service providers realize this, and are also trying to package additional value added services, such as Internet access, energy monitoring and management, cable TV, consolidated billing, etc., to expand their host of services.[8]

ENERGY AND THE WEB

The near-term future for retail choice is definitely blurred at the moment. Stranded cost and operational issues must be resolved for any semblance of competition to develop. Once slamming is relegated to a non-issue, or given the same consideration as slamming laws already on the books, customers will be able to shop for energy services over the Web, just as they do today for thousands of other products and services.

The Web will be the "Energy Mall" for buyers and sellers to engage in all the various energy products and service choices available today and more yet to come. The Web's ability to lower transaction costs for both the buyer and the seller will accelerate this outcome even faster.

The timing of complete electric and natural gas industry unbundling may actually be accelerated due to the advances in Web technology and the growth in the number of homes purchasing low cost PC's and using to the Web to meet their daily products and services needs. Leading edge companies are already offering energy monitoring services over the Web to manage and reduce energy consumption.

The companies, both regulated and unregulated, that hope to gain and prosper in this new energy marketplace will definitely be those that engage the Web to provide their customers with easy to use, efficient, customized and customer focused value added services.

REFERENCES

[1]Smith, Dennis, "Internet Becoming an Efficient Tool for Customer Choice Pilots", *Customer Communications and Metering*, Chartwell Publications, January 1998, Vol. 4, no. 1.

[2]Molzan, David E., and Roger Falcione, "The Pioneer Valley Residential Unbundling Pilot On the Web", Paper presented at the seventh DA/DSM Distributech Conference, Utility Information Technology—Competitive Business Solutions—Customer Satisfaction, San Diego, California, Paper No. VFD. 2 970109, PennWell Publications, January 27-30, 1997.

[3]Nemec, Richard, "Californians Loath to Pick Power Suppliers; Less Than 1% Switch", *Utility Spotlight,* May 5, 1998.

[4]Itron, Inc. Press Release, "Itron Announces Contract With Pacific Gas and Electric Company to Provide Meter Data Collection Services", Business Wire, June 11, 1998.

[5]Feather, John, and Richard Sasdi, "Customer Driven Processes Provide Utilities with Bridge to Deregulation", *PMA Online Magazine,* May 5, 1998.

[6]Tollefson, Phil, Opinion: "Utilities Should Be Deregulated at State Level", *The Gazette,* Colorado Springs, Colo., Knight Ridder/Tribune Business News, June 2, 1998.

[7]Spiewak, Scott, "The Undermining of Retail Access", *PMA Online Magazine,* March 1998.

[8]Aberdeen Group Press Release, "Information Technology Powering Change in the Utilities Business: New Aberdeen Report Analyzes It's Role in Facilitating the Deregulation of Electricity Generation and Supply," Business Wire, May 12, 1998.

INDEX